D1878955

DYNAMICS
OF
SETTLEMENT ARCHAEOLOGY
(Haryana)

DYNAMICS OF SETTLEMENT ARCHAEOLOGY
(Haryana)

DR. R.C. THAKRAN

GYAN PUBLISHING HOUSE
NEW DELHI-110002

DYNAMICS OF SETTLEMENT ARCHAEOLOGY (HARYANA)
(*Archaeology*)

ISBN : 81-212-0708-0

© Dr. R.C. THAKRAN

All rights reserved. No part of this book may be reproduced in any manner without written permission.

Published in 2000 in India by
Gyan Publishing House,
5, Ansari Road, Darya Ganj
New Delhi-110 002

Phone: 3282060, 3261060 Fax: (011)-3285914
E-mail: gyanbook@del2.vsnl.net.in
Visit Our Web Site At http://www.gyanbooks.com

Laser Typesetting at A.S. Printers, Delhi

Printed at Mehra Offset Press, Delhi

Dedicated

to

my dear grandmother,

late Mrs. Shruti

CONTENTS

FOREWORD

Exploratory surveys are an essential, preliminary part of any archaeological work, through which the trained eyes of the archaeologist assess the potentiality of a 'site'. The meaning of exploration has however undergone considerable change over time. Archaeologists no longer simply collect specimens of atefact from a 'site' or look only for what is spectacular from the point of view of enriching museum displays. Several ways in which archaeological explorations are conducted now may be mentioned in order to bring out the concerns which relate to the method of exploration. The ancient is now more on locating and 'documenting' all traceable sites and not simply on collecting samples from a 'micro'-locality. These provide the necessary database for moving on to analyzing the distribution pattern of the sites in the context of the selected 'space', the unit of selected 'space' usually being a modern district. Explorations undertaken at this level, besides generating data on sizes of sites and the nature of culture/cultures present in them, make possible the working out the pattern of nucleation or otherwise, hierarchy among sites and frequency of distribution in terms of regularity or irregularity in distances between sites. All these points relate vitally to the dynamics of the settlement history of an area and thereby of its cultural evolution.

Obviously, there are major limitations to the simple surveys of this kind both with regard to the choice of the spatial unit and in properly ascertaining the succession of settlement-profiles of multi-culture sites. Sometimes, unwarranted generalizations are also seen to emanate from such surveys. Even so, there does not seem to be an alternative to the intensive surveys of the kind that Dr. R.C. Thakran–and others–have undertaken, it is through comparisons of similar surveys, not only in Dr. Thakran's area of study but in other locations as well, that further refinements in strategies of exploration will emerge.

Based on intensive fieldwork in the district of Sonipat in the state of Haryana, Dr. Thakran's work, which is a welcome addition to the existing, rather inadequate, literature of the genre of settlement archaeology, contains valuable material for understanding the spatial pattern of the evolution of material culture at a micro-level, nevertheless with the potential for making fruitful inferences for a larger canvas. The presentation of the material has been made effective with maps and diagrams, providing instant impressions of the patterns worked out in the text. Dr. Thakran has already published regularly on the archaeology of the Indo-Gangetic divide; his work should be a reminder to us that frequently made facile generalizations about India of the past as a whole need to be coustansly checked with the details that the country's various segments reveal through meticulous fieldwork.

B.D. Chattopadhyaya

PREFACE

This work, *Dynamics of Settlement Archaeology* is on *Sonipat District of Haryana* and is addressed to a specific problem of Ancient Indian History. It primarily deals with the emergence and evolution of settlements from the early Harappan to the early historical period in Sonipat district of Haryana. Since, it is addressed to a particular problem on settlement patterns in a district of Haryana Settlement is perceived as a process of social activities undertaken to meet basic biological and cultural needs of human beings. Obviously these activities take place within the context of social and environmental units. Social units comprise household, local group and other sectors of society, which are related to dwellings, settlement and settlement patterns. Environment involves the study of ecology, landscape, vegetation, etc.

The main source material is provided by explorations either previously conducted by others or by me. In addition both excavated and explored material from adjoining areas and ethnographic information is also an important source material for the present work.

The problem of settlement patterns has been viewed from different angles. Aspects like ecology, topography, rainfall, hydrology, landuse patterns, movement and migration or peoples, technology, social organization and material remains form the central hub of this analysis.

The main endeavour has been to ascertain inception of the process of habitation and also to trace its sequence and spread over the period. Changes were bound to surface in this process. These changes and their catalysts, such as natural and social factors have been central to our understanding of interconnections. To achieve these objectives crucial information on soil formation, drainage, hydrology, sources of water, vegetation and subsistence activities was obtained during an intensive combing of the area under review.

This study, however, like any other study suffers from its own inherent limitations:

(i) One of these limitations is the total lack of excavated material from the area of discussion.

(ii) The general situation of excavations in the immediately surrounding areas is also not very satisfactory, because most of the excavated sites have been opened up to a very small extent to ascertain the vertical sequence.

(iii) Reports of the excavations have not, yet, been published in most cases. This is a serious lacuna with regard to unpublished reports of the excavations, especially of early historical sites.

(iv) Shortage of data on early medieval settlement in and around the area of study is startling.

(v) Non-availability of data on past subsistence practices.

The present work has been divided into five chapters followed by a conclusion and appendices in the form of a list of single or major culture representative sites and a short note on subsistence patterns and finally a select bibliography.

Chapter I, is the introduction to our 'region'. It deals with the location, name, natural resources, administrative history and modern settlements and subsistence systems along with a brief resume of past archaeological activities in the State.

Chapter II, is on my 'Explorations' and mainly describes the explored sites, their physical condition, cultural sequence and location in relation to their respective natural background.

Chapter III, deals with the distribution of settlements. It gives a brief outline of previous studies, culturewise distribution of settlements, cultural sequence, their respective time span, areas of frequency, location and lines of extension of settlements. This also explains the involved processes of site selection.

Chapter IV, deals with Settlement Patterns and has a detailed analysis of the emerging settlement patterns and their catalysts. For this purpose the area under review has been divided into working sub zones based on the administrative units and physiographic features. The existing patterns in rainfall, hydrology and land-use on the one hand and the past patterns in social organization and technology on the other help provide an understanding of the dynamics of their intrplay to promote settlement patterns. This offers an insight into the past cross-currents of operative ideas and also exemplifies as to how distinct patterns were shaped in different zones and periods.

Chapter V, is a study of Material Remains. In this Chapter concentration is based on the artifacts representing the said stages of development. The chronological discussion of the material brings out the processes of development of society. This is corroborated by parallels with reported material from other settlements in the surrounding areas.

ACKNOWLEDGEMENTS

This is extremely pleasant to express my deep indebtedness to Professors P. Bhatia, R.S. Sharma, R.L. Shukla, A.K. Sinha, B. Mukherjee, D.N. Jha, K.M. Shrimali, D.K. Bhattacharya, S.K. Vij, T.K. Venkatasubramanian, Drs. B.P. Sahua, S.Z.H. Jafri, R.P. Rana, M.L. Bhatia, S. Bhowmik, S.S. Rana, P.C. Patanjali, R.K. Chattopadhyay and Shireen Ratnagar of University of Delhi, Professor B.D. Chattopadhyaya of JNU, Prof. V.N. Misra of Deccan College, Poona, Prof. Suraj Bhan, Prof. U.V. Singh, Dr. M. Singh Ch. Dalip Singh of Kurukshetra University, late Prof. S.R. Phogat, Dr. Manmohan Sharma, Dr. A. Singh of Maharshi Dayanand University, Rohtak, Sh. A.K. Sinha, Sh. S.C. Saran, Sh. R.S. Rana of Archaeological Survey of India and Sh. Satish K. Dahiya and Nagender Sharma for their multifarious help in this work.

I do wish to express my gratefulness to the people and officials of Sonipat district, all the librarians and staff members of Archaeological Survey of Inda, National Museum, Central Secretariate Library, ICHR Library, AIIMS Library, Central Ground Water Board Library—all in New Delhi, Central and Rattan Tata Libraries of University of Delhi, Geological Survey of India, Faridabad, Deccan College, Poona, Kurukshetra University, Kurkshetra for their cooperation and help in providing essential material. I also thank Sh. Raghbir Singh Dahiya of Rattan Tata Library for arranging necessary material. I acknowledge receipt of financial assistance from the UGC for this work.

I thank Sh. T.K. Roy, Sh. Krishan Lal, Sh. O.D. Tyagi, Sh. Vinod Kumar Bhardwaj for maps and photographs, and Sh. R.K. Gupta for prompt typing.

I do not know how to express my indebtedness to my wife and members of my family for their unflinching support and endurance. Finally, I alone am responsible for all shortcomings.

R.C. Thakran

ABBREVIATIONS

ABORI	*Annals of the Bhandarkar Oriental Research Institute*
Am An.	*American Anthropologist,* Chicago
AI	*Ancient India,* New Delhi
Anthropology	*Anthropology,* New Haven
Antiquity	*Antiquity*, London
BNIS	*Bulletin of National Institute of Science*
CF	*Cultural Forum*
CII	*Corpus Inscriptionum Indicarum*
EI	*Epigraphia Indica*
EW	*East and West*
GJ	*Geographical Journal*, London
GSI	*Geographical Survey of India* Publication
HSHC	*Haryana Studies in History and Culture,* Kurukshetra
IA	*Indian Antiquary*
IAR	*Indian Archaeology-A Review*, New Delhi
IHR	*The Indian Historical Review,* New Delhi
IJHS	*Indian Journal of History of Science*
IMB	*Indian Museum Bulletin*, Calcutta
ISPP	*Indian Studies Past and Present*
JASB	*Journal of Asiatic Society of Bengal*, Calcutta
JESHO	*Journal of Economic and Social History of the Orient*
JHS	*Journal of Haryana Studies*, Kurukshetra

JISOA	*Journal of Indian Society of Oriental Art*
JNSI	*Journal of Numismatic Society of India,* Varanasi
JRAS	*Journal of the Royal Asiatic Society*
URJ	*Kurukshetra University Research Journal,* Kurukshetra
LK	*Lalit Kala*
ME	*Man and Environment,* Poona
MI	*Man in India,* Ranchi
NC	*Numismatic Chronicles*
PIHC	*Proceedings of Indian History Congress,* Delhi
PPHC	*Proceedings of Punjab History Conference,* Amritsar
PPS	*Proceedings of Protohistoric Society*
Puratattva	*Puratattva,* New Delhi
QRHS	*Quarterly Review of Historical Studies,* Calcutta
QSR	*Quarternary Science Review*
S. Am	*Scientific American,* New York
VIJ	*Vishvesvaranand Indological Journal,* Hoshiarpur
WA	*World Archaelogy*

LIST OF ILLUSTRATIONS

LIST OF MAPS

1

INTRODUCTION: GEOGRAPHICAL BACKGROUND

This work covers Sonipat District of Haryana State, which appeared on the map in 1972, vide Haryana Government Notification No. 6050-E(IV)-72/45724, dated 22nd December, 1972.[1] It stands in the south-eastern part of Haryana State within 28°48' 30" and 29°11' 54" north latitude and 70°28' 30" and 77°13' 40" east longitude. It falls in Survey of India Toposheets number 53-C, D, G and H encompassing an area of 2206 square kilometre.[2] It is bounded by district Karnal (now Panipat) and Jind in the north and north-west respectively. In the west and south-west it is bounded by district Rohtak. The State of Delhi covers it from southern side. And the Yamuna, the sole perennial river of this region, forms the only natural dividing line between the district Sonipat of Haryana in the west and Meerut district of western Uttar Pradesh in the east (map I).

So far as the administrative arrangement is concerned it is divided into two main tahsils, Sonipat and Gohana, with their respective headquarters, for purpose of administrative convenience. These are further divided into smaller units known as Community Development Blocks for smooth running of administration. There are seven such blocks by the name Sonipat, Rai, Kharkhauda, Gohana, Ganaur, Kathura and Mundlana. Each block consists of the following number of villages Sonipat - 79, Ganaur - 78, Rai - 63, Kharkhauda - 38, Gohana - 35, Mundlana - 34 and Kathura - 20.[3] Thus, Sonipat district comprises 347 villages. Of them 17 are uninhabited at present.

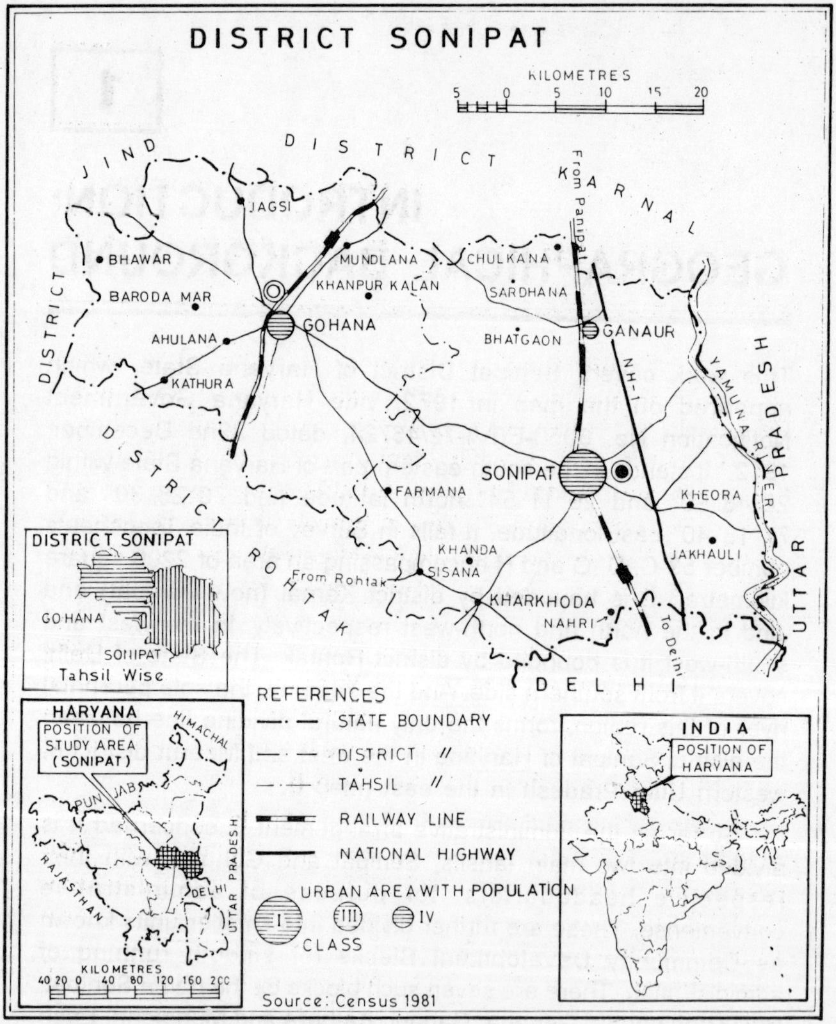

Map. 1 : District Sonipat

Population Size

The district accommodates 846,765 persons[4], which is 6.57 per cent of the total state population. The density of population comes to 384 persons per square kilometre in the district as a whole as against 292 persons of the state. Population-wise the district ranks second among the twelve districts of Haryana. With regard to tahsil-wise distribution of population Sonipat tahsil claims first position by harbouring the highest number of 435,538 inhabitants as against 324,213 persons of Gohana tahsil. The density of population is also higher in Sonipat supporting 452 persons per square kilometre, whereas in Gohana tahsil it is 272 as against 384 of the district.[5] Regarding the distribution of population at the village level in the two tahsils it can be stated that by and large the villages in Gohana tahsil have an average population of over 2000, whereas, in Sonipat it is about 1200 persons. That is why villages in Gohana tahsil are scattered, but of reasonably large size, whereas, in Sonipat these are in close proximity, but smaller. These differences between the two tahsils are very helpful to understand the interplay of forces of production. In spite of a gradually growing shift of population from rural to urban centres the pattern of distribution between rural and urban has not changed much over the years because, even, today 90.2 per cent of the population inhabits rural sector.[6]

Antiquity and Administrative History of Sonipat

Sonipat town, the district and tahsil headquarters, is (29° 00' N 77° 1' E) about 40 kilometre north of Delhi and 48 kilometre north-east[7] Rohtak (map 2).

It seems the city of Sonipat was founded by Raja Soni, supposed to be 13th in descent from Arjuna, brother of Yudhisthira.[8] Sonaprastha, identified with Sonipat, has also been referred to by Panini.[9] This is further supported by archaeological evidences as NBPW and its associated artifacts were collected from Sonipat by B.B. Lal.[10] Occurrence of a large hoard of 1200 hemi-drachms of Graeco-Bactrians[11] and about 10,000 coins of Yaudheyas, Kushanas and later Kushanas speak of its significance in early historical and historical periods.[12] This is further supported by a copper plate-seal of Harsha Vardhan from this place. This seal provides the genealogy of the Pushpabhuti rulers of Thanesar.

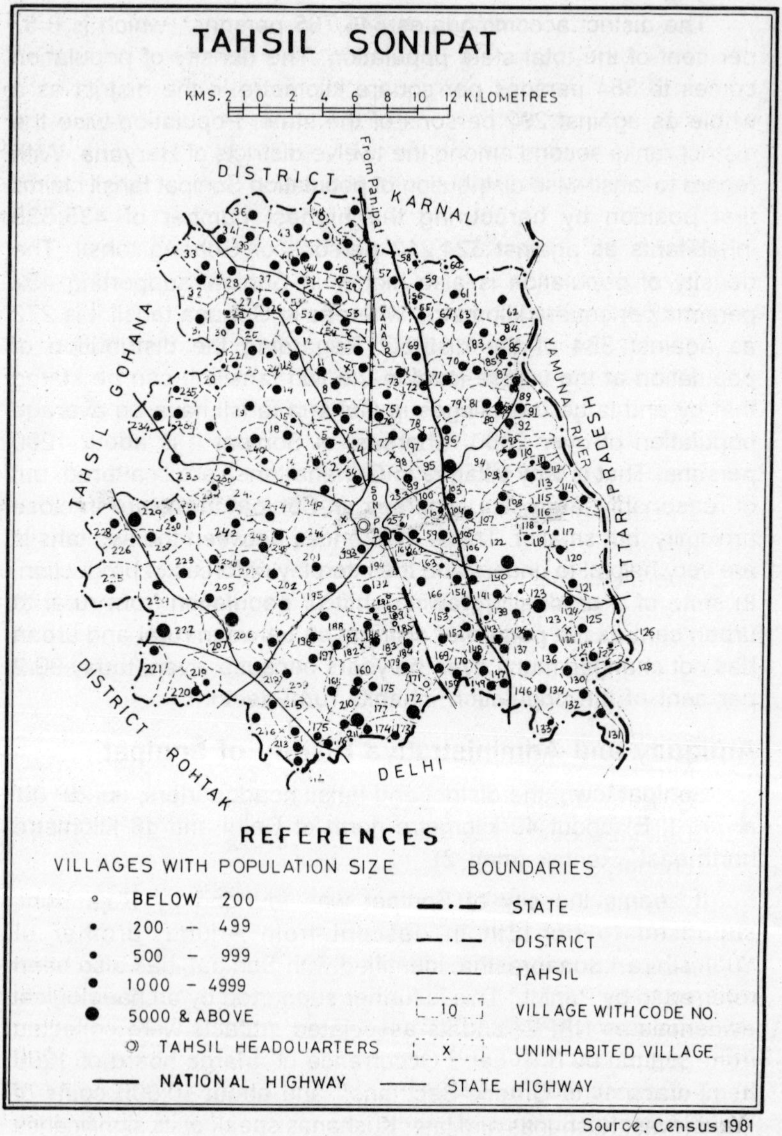

Source : Census 1981

Map. 2: Tahsil Sonipat

The town of Sonipat seems to have been located at a very significant (politically and economically) strategic point.[13] It was connected by a north-south trade route running almost parallel to the Grand Trunk Road between Mathura and Thanesar in the early historical times. It was also connected by an east-west route between the Indo-Gangetic Divide and to the Ganga-Yamuna Doab. Sonipat and Baghpat happened to represent two nodal points along this route on either side of the Yamuna. Similarly Sonipat lay on the subcontinental route from Jullundhur to Tamralipti. This route was given proper shape by Ashoka and later by Shershah Suri.[14] Subsequently this region was included in the empire of Mihir Bhoja – I (AD 836-89) of Gurjar Pratihar dynasty as stated by Pehowa inscription. And thereafter, it is said to have remained under the Tomar rulers of Delhi who ruled over the whole of Haryana.[15] The word 'Haryana' first occurs in a Delhi Museum inscription dated AD 1328.[16] The area is described as heaven on earth and was included in Delhi.

During the early medieval period, in AD 1037, Sultan Mahmud of Gazani defeated Dipalhar, the governor of Sonipat. In this operation Masud, the son of Mahmud, is said to have been proceeded from Hansi to Sonipat, possibly through Khokhrakot (Rohtak) in order to extend his suzerainty over these areas. Subsequently the Tomars succeeded in recovering these areas from the Gazanavids who ruled over these areas as indicated by another inscription dated AD 1328, now housed in Delhi Red Fort Museum.[17] But the situation changed when the Gauris and Chauhans replaced the Gazanavids and Tomars at Lahore and Delhi respectively by the middle of the 12th century. This political change ensured a keen contest between these two powers of the day to gain supremacy over Delhi and its environs. In this process a famous decisive battle of Train or Traori, a place about 21 kilometre south-east of Thanesar in Kurukshetra district, was fought in AD 1192, which established the Gazanavids at Delhi including Sonipat.[18] Subsequently Sonipat finds expression in one of the inscriptions of Mamluk dated AH 676 (AD 1277), which was inscribed by Ghiyas-ud-din Balban wherein the name of the Muqta of Sonipat is mentioned 'QirBek'.[19] This has further been supported by the mosque of Abdullah Nasir-ud-din built in AD 1271, in Balban's reign in memory of the celebrated saint as mentioned in the arch inscription on the outside of the mosque.[20]

Being in close proximity to Delhi, Sonipat seems to have remained continuously under the hegemony of the rulers at Delhi as revealed further by the tomb of Khawaja Khizar at Sonipat. This was built in the reign of Ibrahim Lodi in the memory of another local celebrity. It is a very fine structure of stone and equally significant from the point of view of the history of architecture.[21] Likewise a square pillar located near the graveyard of Imam erected in AD 1479, during the regime of Bahalol Lodi points in the same direction.[22]

Sonipat maintained its strategic position, even during Mughal period. In the time of Akbar when his minister Todarmal divided north India into smaller administrative units, popularly known as circles, the area of Sonipat constituted parts of the *Subah* of Delhi. This tract of land was usually granted in military *Jagirs* to the nobles by the Mughal emperors.[23]

After the fall of the Mughal empire the claims of the imperial grantees were contested by various personalities, but the Sindhia managed to conquer this area. However, he was not destined to hold for long to the west of the Yamuna. Because by the treaty of 'Surji Arjungaon' of December 30, 1802, the western possessions of Sindhia passed to the British and came under the administration of the North-Western Provinces. And the British were not inclined to hold large territories to the west of the Yamuna for their own problems and limitations. As a result a number of chiefs, who had either supported the British or remained neutral *vis-a-vis* the Marathas, were settled in this tract to form a barrier of independent outposts between the British border and the Sikh State of Maharaja Ranjit Singh. Accordingly Kharkhauda and Sonipat were assigned to Lal Singh of Kaithal as life *jagirs*. But this estate lapsed to the British Government on the death of Lal Singh in AD 1818. In AD 1824, five districts namely, Panipat, Rohtak, Hissar, Gurgaon and Delhi were carved out as separate units. And consequently the whole area of Sonipat along with its surroundings was brought under the resident at Delhi and this arrangement continued till AD 1832, when the resident became Commissioner. And Sonipat was passed on to district Rohtak and remained there until AD 1841. In that year the Rohtak district was abolished and Sonipat area once again transferred to Delhi. But in the very next year—AD 1842, Rohtak district was reconstituted and Sonipat also became its part. Subsequently after AD 1857, the Delhi territory along with the Yamuna belt upto Karnal including Sonipat, was transferred to North-West Provinces of the Punjab by the Government Notification No. 606 of the 13th April, 1858.[24] But again in 1861, Sonipat was affiliated to Delhi to keep a close vigil on the activities of residents of villages like Kundli, Samri and Purkhas, all in Sonipat, as they participated in the 1857, freedom struggle against the British imperial. power.[25] 1862, onward till 12th December, 1911, when the capital changed from Calcutta to Delhi and Delhi tahsil was separated from Punjab, Sonipat remained with Delhi. But in the very next year of 1912, from Ist October, Sonipat was transferred to Rohtak.[26] And thus, remained with Rohtak until creation of separate district of Sonipat in September 1972.[27]

Antiquity and Administrative History of Gohana

Gohana, the modern town and tahsil headquarters, 29°8' N 76°42' E (map 3) is situated on the northern extremity of a great natural depression extending about 75 kilometre southwards.[28] The earliest most reference to Gohana is 'Gavam Bhawam' in the *Van Parva* of the Mahabharata.[29] There are two popular versions for the origin of the name of the town based on *Van Parva* reference. According to first Gohana is composed of two words—'gau' and 'hani', the former means 'cow' and the latter means 'loss'. It means the loss of cows or cows were lost here. The second version is also composed of two words, 'gau' and dahan', the latter meaning a small water course in the local dialect. In the local parlance this term also stands for 'cow-wealth' and the word used for it is 'gaudhan'. This points to the assembling of cows near or around the large depression. Here in both the versions cow is a common mention suggesting the importance of cow. Cow seems to have been a source of livelihood as it was during the Vedic period. Pastoralism seems to have been a major constituent of subsistence activities and a similar situation existed till very recent past in this zone. In almost every household there used to have been a large number of cattle. The economic condition and social status was adjudged with the heads of cattle kept by an individual. And the importance was so high that one male person was assigned for this job. This was because the ecological conditions were most suiting to this animal and this type of mode of subsistence. The mention of 'gawam bhawam'

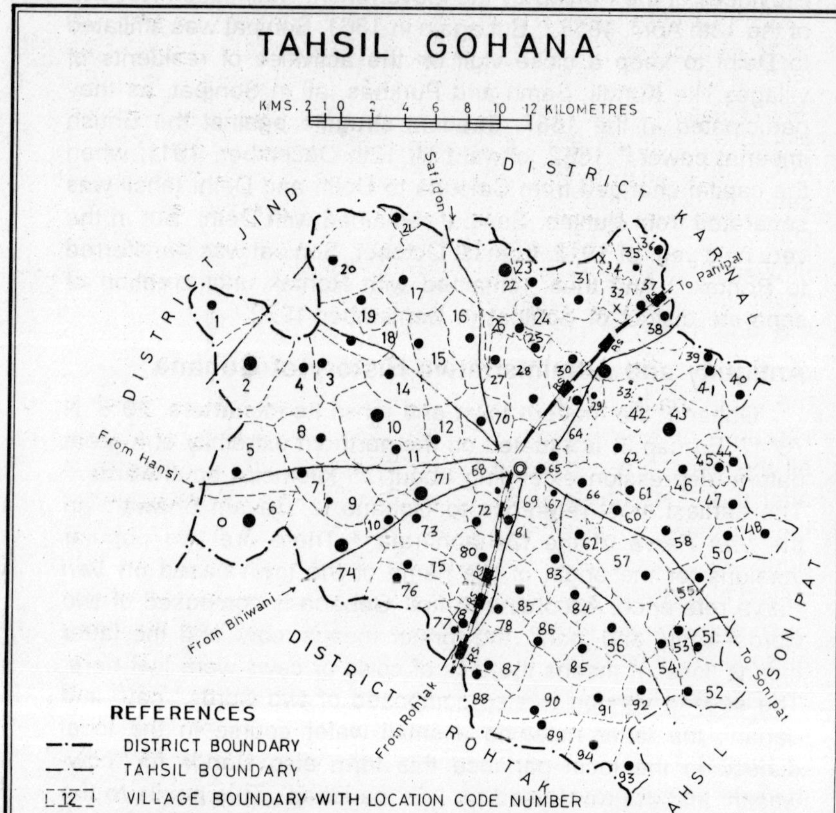

Map. 3: Tahsil Gohana

is suggestive of a house of cows on a place where cows were generally kept or where cows were generally bred. It is said that cows were assembled on the mound where present Gohana is located till AD 1238[30]. It is customary in this area to assemble cows at a particular commonplace in the morning and later to take them to the pasture ground for grazing.[31] Thus, cow and commonplace is significant in this respect. It is also said that this place was called as 'gowdhan'—'cow-wealth', which later corrupted to Gohana.[32] Thus, in the light of the above references it can safely be concluded that Gohana and its surroundings were most conducive to pastoral activities, since ample surface water in the huge depression, grazing grounds and moderate to dense vegetal cover were available the year round. This is also suggestive of the long history of Gohana.

Gohana is said to have had four gates leading to Rohtak in the south, Delhi in the south-east, Panipat in the north-east and Hissar in the west suggesting its links via road to these places in all the directions. There was also a tank by the name Rohtas after the name of the king of Rohtak.[33] This tank is not visible, though there is certainly a huge depression, especially to the north-west of Gohana, which receives rain water from Karnal district and assumes considerable proportions during rainy season. Such depressions provide water, shade, grasses to the cattle in all seasons. It has special significance to people rearing animals.

Regarding the historical and medieval period of Gohana it can safely be said that its fate was depended upon either Sonipat or Rohtak or Jind as the situation arose from time to time as there is no separate and independent reference to Gohana. It was, however, assigned to Raja Bhag Singh of Jind for his neutrality or support to the British. The British adopted this strategy to create a buffer or protective zone between the Sikh state of Maharaja Ranjit Singh and the British border for their own strategic calculations. But after the death of Bhag Singh in AD 1820, this area lapsed to the British. In AD 1824, with the creation of Rohtak district Gohana and Kharkhauda tahsils were passed on to Rohtak and the area remained under the residents of Delhi till AD 1832. The district was ultimately liquidated in AD 1841. Consequently Gohana went to Panipat district. But, in the very next year in 1842, the Rohtak district was carved out again by detaching along with divisions of Delhi and Hissar from North-

Western Provinces. And after 1857, this was transferred to Punjab on 13th April, 1858. And Rohtak district along with its four tahsils of Gohana, Jhajjar, Sampla and Rohtak was attached with Hissar division. It remained with that division until the year 1884, when eventually Hissar division was abolished and Rohtak along with its tahsils was transferred to Delhi division. And subsequently in 1912, Rohtak district was created along with Gohana and thus, it remained with district Rohtak[34] till the creation of district Sonipat in September 1972. Now it forms apart of Sonipat district.[35]

Physiography of District Sonipat

The Sonipat district is situated on the watershed dividing two mighty river systems—the Ganga draining into the Bay of Bengal and the Indus falling into the Arabean sea via Sind. These alluvial plains overlies the rocks of Algokian system and are consisting of interfringing beds of clay, sands of various grades, silts intermixed with *Kankar* in varying proportions are underlain by the older and heavier alluvium of the Indo-Gangetic—the quarternary-alluvium. These range in age from the upper Pleistocene to Holocene or recent age.[36]

There is a tremendous influence of the Himalaya and their mighty rivers on the physiography of this area. As the area is a part of the vast Indo-Gangetic plain, it is alluvial in nature and devoid of any prominent topographical features, in general appearing monotonously flat to a layman. But intensive studies on large scale would reveal significant variations in the surface configurations to one interested in micro-regional geography. The meanderings of the river Yamuna and other contemporary water courses through the millennia have left their deep imprint on the physiography of this area. A general gentle sloping tendency of the land to the south/south-east from the north/north-west is evidenced by the main direction of the Yamuna from north to south. There is a general fall from north to south with about 30-35 centimetre per kilometre, especially in the Yamuna flood plain[37] as the plains are 760 feet above sea-level in the north and 690 feet above sea-level in the south-eastern parts of the district near the State of Delhi.[38] The district can physiographically be divided into two main distinct localities—the Khadar (new alluvium) and the Bhangar (old alluvium). It will be reasonable to elaborate at length these units separately in the following lines.

The Khadar

The Khadar is a low-lying flood plain of the river Yamuna in the eastern section of the district. It extends to about 10-15 kilometre in the west in a serpentire fashion, which has been marked out during the intensine survey of this area. A thick dividing line separates the two contrasting zones. And the map also demonstrates the respective expanse of the two zones. The riverain plain is 20-30 feet lower than the adjoining upland plain in the west. This is having a sloping tendency towards the south/ south-east corresponding with the Yamuna course. The Khadar is characterized by unconsolidated new alluvium, which generally occurs along the Yamuna course. The soil is composed of sand, silt and clay proportions, which is light coloured and poor in calcareous matters consisting of lenticular beds of sand, clay gravel and sand from the main aquifer zones.[39] The Khadar soil has a submature profile with domination of clay concretion, fine clayey loam left by the receding floods. Here salt percentage is very low, except low-lying, waterlogged and elevated areas untouched by the river water.[40] Generally soil of this area used to remain wet for most part of the year, before the intensive exploitation of soil water began with the introduction of tube-wells, paddy cultivation and earthen bund along the right bank of the Yamuna to control floods, since the water level remained very high only 5-10 feet below the surface. For this reason the area hardly suffered from real dry conditions.[41]

The Khadar appears fairly flat, low-lying area except the ridges of sand along the river, but the discerning eyes can recognize important lines of demarcation breaking the monotony of the Yamuna plains.[42] Such a scenario is caused by variations in surface configurations, though of minor nature, yet, very potential to support a good deal of biodiversity.

The Bhangar

The Bhangar is an upland plain consisting of old alluvium of early Pleistocene period and lies immediately to the west of the Khadar. It is parallel to the Khadar in its entire north-south length. The Western Yamuna Canal at least in this district, almost marks along the dividing line of the Khadar and provides irrigation to the tract seriously ridden with the problem of salinity.[43]

Considering from relief point of view the plains do not appear uniformly level area, but relatively flat. The slope of this area is insidiously advancing to the south-east from the north, which is intercepted in between by local slopes contradicting the general inclination tendency. The height of the plain in the north of the district is 760 feet above sea-level, which comes down gradually to 700 feet by the time the southern edge of the district appears near Sonipat and Delhi border, with the average fall of 30-33 centimetre per square kilometre being similar to that of the Khadar.[44]

The area also has some low, but long ridges, forming a distinct feature of this zone, running almost north-south. These are rising about 30-40 feet, which perhaps are the remnants of the northern extension of Aravallis in Gohana region.[45] These are broken in parts and dissected by small and seasonal shallow low-lying water courses. Of these sand-hills, the eastern most keeps cropping up within a few kilometre of the Delhi border in the west. The central ridge runs right through Rohtak itself and the western borders the Hissar district. Interestingly these are at right angles to the direction of prevailing winds. Similarly sand dunes also occur with north-south orientation.[46]

The older alluvium is overlain by thin layers of sand at places where the wind is an active agent of transportations and deposition,[47] especially in summer when sand storms or squalls occur very frequently in the months of May and June. It is not uniformly deposited. Its coarseness is suggestive of its aeolian origin from Rajasthan. It seems the process of sedimentation was not uniformly consistent for the occurrence of conglomerate horizons at different levels suggests the intervening periods of rest. Pans of carbonate (*kankar*) occur within 3 metre of the surface above water-level.[48]

Soils are usually good, sandy, clay loam, sandy loam and alluvium. Locally these soils are termed in different manner. The sandy is known as 'raitile' or 'teeba', sandy loam as 'bhur', loam as 'rausli', clay loam as 'karde' and clay as 'dakar'/'chikni'. The last one splits into fissures after irrigation and is very difficult to till after it has dried out. This exists only in depressions as the greater part of the argillaceous matter is washed down by rains from the surrounding higher lands. These can also be seen along the drainage lines known as 'dahari'.[49] The pH of these soils varied from 7.5 to 9.0, while the organic matter runs in between 0.75 to 0.2 point, which is generally found in the upper black layer known as humus. This tends to bind the particles together, reduces pore size and separates the soil material into clumps. This allows better aeration, increased percolation and easier root penetration. Such soils are deficient in nitrogen, contain medium quantity of phosphorus, but are rich in potassium. It is pointed out that phosphorus is even less than nitrogen, since the former is only 1.01 per cent while the latter is 1.13 per cent. And potash is about 3 per cent.[50]

Saline Soils

Salinity problem is not alien to this area for such soils do occur in between. Some of the most affected villages are Busana and Sarsad in Gohana tahsil.[51] And generally saline soil pockets are witnessed at regular intervals over the landscape resulted both by human and natural forces.

Hydrological Background

The underground water system forms an interesting study, since the district is away from the high rainfall zone of the Siwalik hills and traversed by a sole perennial river only on its eastern boundary. Being semi-arid in nature, experiencing irregular and uneven distribution pattern of rainfall and having shallow to medium depth of water-level with the existing higher regime of underground salt solution and the nearness to the perennial and seasonal streams becomes still more interesting. And the quality of subsoil water is, most likely, affected to a greater extent by these factors. Therefore, the quality is, on an average, saline and brackish excluding the riverain plain of the Yamuna in the east. Apparently on this basis the district can be divided into two major subsoil water zones almost in tune with the geographical classification—the fresh water zone of the *Khadar* 'in the east' and the saline zone of the Bhangar 'in the west'.[52] Keeping in view the existing hydrological regime still more micro zones cannot be ruled out at this stage, which will be taken care of in the subsequent chapters. However, presently the main discussion revolves around the existing main divisions.

Fresh Water Zone

The sweet water belt lies along the Yamuna comprising areas between the Western Yamuna Canal in the west and the Yamuna in the east. Although this belt does not present uniform water-level of fresh water as some pockets with saline and brackish water are marked in between. Yet, overwhelmingly this zone possess a good quality of water. This has been resulted in due to the consistent river action for centuries now and most of the salt contents in the soil have either been washed away completely or reduced considerably depending upon the consistency and intensity of flood waters of the Yamuna. As a result the water in this belt is sweet and was available in abundance at shallow depths varying from 4 to 15 feet.[53] But more recently the water level has gone down considerably.

The natural resource in the form of subsoil water was successfully being exploited intensively by the farmers through tube-wells for the past almost two decades, without lowering the water-level for high refreshing rate due to rains and annual inundation. But more recently the ongoing continuous lean phase of below normal rainfall is causing concern as resulted in the falling of the water-level growingly. This has further been accelerated by stoppage of the Yamuna flood waters to this area after enlargement of the earthen bund from north to south all along the western bank of the Yamuna from Babail, about 8 kilometre north-east of Panipat, in Panipat as a protective measure. As a result this checked altogether the natural regenerative mechanism leading to considerable fall in subsoil water-level in the area. Furthermore, non-existent canal network is promoting the deepening of the already ever-sinking water-level. And last, but not the least the increasing over-exploitation of the ground water with the shifting crop emphasis from sugarcane to paddy and wheat combination expedited the process further. This change was resulted into by a general awareness and availability of intensive farming techniques with the introduction of 'green revolution'. However, if any arrangement is not made to arrest this ongoing deepening process of water-level in the foreseable future the situation is likely to deteriorate beyond economic redemption reach of the farmers in particular and the common men in general.

In this zone the ground water gradient is from north to south, a course more or less parallel to the Yamuna river. The maximum elevation of water table is 230 metre above mean sea-level in the north and north-west and dips to the minimum of 207 metre in the south.[54]

Saline Water Zone

The western parts of Sonipat tahsil and the entire Gohana tahsil fall in the brackish water belt with exception of some patches where the quality of water is a little good, but only in shallow aquifers resulted from constant visiting of canal water for irrigation purposes and high rate of monsoon recharge in the fifties and sixties. A similar trend is witnessed near natural and artificial depressions where rain water is stored in sufficient amount to maintain and accelerate the rate of recharging of subsoil water. The maximum depth of this water sheet is hardly 30-50 feet, not promising enough,[55] to withstand the over exploitation in absence of any consistent and assured refreshing arrangement.

In this saline zone there is a deep seat of underground brackish water. Salt dissemination can also be seen in the surface soil of this entire section. The general water-level is between 15-20 feet,[56] though a rising tendency was witnessed in the recent past, which was ultimately frustrated by brick-lining of canals blocking seepage, effective drainage of the area, increased withdrawal from ground water reservoir than the refreshing rate, low rainfall and precipitation for the last several years. In the south-western parts the water-level is still low around 20-30 feet. And for the aforesaid measures to check waterlogging conditions, schemes of water conservation, over exploitation of subsoil fresh water through tube-wells for irrigation and continued receding rainfall pattern for several years now there appears a danger of drying up or exhausting the underground reservoir. If this is not managed to reverse somehow with the help of time tested traditional water harvesting practices, being ignored in the name of modernisation, the ultimate threat to the agriculture and the people is imminent.[57]

Drainage Pattern

The Delhi branch of the Western Yamuna Canal, flowing over the crest of the watershed, forms the main dividing line between the two separate drainage systems. To the east lies the Yamuna *Khadar*, where the Yamuna provides the only suitable natural outfall for the drains of the district. However, the topography of the district does not permit the natural drains to flow directly into the Yamuna, though the area lying to the east of the Delhi Branch is directly drained into the Yamuna. This area is drained by an artificial channel known as Drain No. 6 coming from Naraina in Panipat tahsil of district Karnal (now Panipat) and becomes co-existent with the so-called old bed of the Yamuna—locally known as 'Budi Nadi' or 'Ganda Nala'.[58]

The area to the west of the Khadar has a master slope from north/north-west to south/south-east and is drained mainly by Drain No. 8 (and diversion Drain No. 8), which is locally called as 'Ganda Nala' around Gohana. It runs through the entire length of the district in almost north-west to south-east direction joining ultimately the Yamuna in the east. In its entire length of about 70 kilometre from Gohana to the Yamuna joined by several smaller drains to benefit the most parts of Gohana and Sonipat tahsils.[59]

Waterlogging Condition

This is a serious problem of this district and worst condition is experienced during rainy season, especially in those areas where water-level is 0.5 feet. And the areas with 5-10 feet sub-soil water-level are suggestive of a danger zone. The rise in water-level leads to appearance of 'thur'—a white/ash coloured material consisting of harmful salts. This material is witnessed subsiding after the rains, but forms a crust over the surface betraying its existence. Under the influence of these salts the soils become unfit for agriculture for soil moisture affects productivity of land. Such soils can support only salt resistant plants.[60]

Climate

The climate of the district is a reflection of the existing sub-tropical continental and semi-arid environmental condition of the northern plains.[61] It is extremely dry and intensely hot during summer and equally cold during winter. It is for most part of the year of the pronounced continental character marked by high variations in temperatures.

Climatically the whole year may be divided into four seasons: summer locally called 'Kharsa' or 'Garmi' begins by month of March and lasts till July, the monsoon (humid) season is called 'Barsat' or 'Chaumasa' extends from July to September, temperate season—a transitional period from monsoon to winter spreading over the months of September to November and finally the winter season 'Jada' begins with December and lasts till February.[62]

In the summer season normally very hot, dry and strong western winds blow over the region, which are locally called 'loo', temperature ranges from 40°C to 47°C. Sometimes it is still higher in exceptional conditions. Squalls and *Andhis* are a common feature of this season. Generally in the afternoon the percentage of humidity is reduced to only 20 per cent. The dry winds and scorching sun are very useful to increase fertility of soil and especially, when it is already tilled. Soil erosion is not a very serious problem in this district except minor effects of fast western winds in the months of May and June and that too generally confined too aeolian rising sandy pockets. To minimize its gravity and check extension of its effects farmers used to grow *sarkanda*, a kind of reed, on the boundaries of their fields in the recent past, which is now becoming increasingly a thing of the past, particularly in context of this district.[63]

During winter season westernly winds sweep the area. Sometimes frost is experienced especially when the mercury falls to 2-3°C during night time, but fortunately during day time the sky is clear on an average and the sun shines brightly providing relief from biting cold. Mediterranean winds are very crucial for the 'rabi' crops. A few showers of light intensity are experienced between December and February because of westernly depressions.[64]

A dry summer and cold winter sandwiches a wet monsoon. The south-westerly and the north-easterly current of summer monsoon, the former being more active than the latter, ease the dry situation. And subsequently this moist air of oceanic origin penetrate the district bringing moderate to heavy rains from late June to July and to early September. This season experience high variation in the rainfall regime causing floods and droughts from time to time as given below. From October onwards the weather remains mainly dry except for a few intermittent showers due to westerly depressions.[65]

Rainfall

The district lies outside the full track of monsoon. The climate of the area is subtropical continental and semi-arid. The rainfall is capricious and often scanty. It generally increases from the south-west to the north-east and reaches its peak in the month of July. That month sees 75 per cent of the annual rainfall. There is a considerable local variation also from year to year, typical of a semi-arid area. The range of variation is demonstrated by the records of two old guages at Gohana and Sonipat set up in 1850.[66]

The semi-arid zone under discussion is, of course, extremely sensitive to minor fluctuations in the prevailing pattern of rainfall and temperature. Positive changes make it a potential conducive zone and the vice-versa for human activities. And in adverse conditions, especially, when people are devoid of matching techniques, tools and human organization to negotiate with, life becomes miserable.

Floods

If the actual rainfall is in excess of the normal by more than twice the mean deviation that year is defined a flood year. And It is interesting to note that the longest number of rainfall abnormalities like floods and droughts also do occur where the general rainfall pattern is erratic and precarious. Since, the whole of Haryana state reel under such climatic conditions, a characteristic of semi-arid zones, there is every likelihood to experience rainfall abnormalities every now and then. A brief account of this is given below to highlight variations in prevailing monsoon realm.

The Khadar area is basically a flood plain of the Yamuna and has its own long history in this respect. A look at the past records would reveal that this area was ravaged by floods, since time immemorial. The tendency of flooding is compounded by the inadequate capacity of the river course to carry the full flood discharge as the flood plain is flat. With the result, whenever, the river is in spate the flood encroaches upon the whole Khadar belt. Adjacent villages used to get marooned and much havoc is caused to life and property. The floods were mainly confined to the area between the river and the upland Bhangar belt. Although the history of floods travels back in the remote ancient past, medieval evidence on floods, among the recorded floods of the near past mention may be made of the following years: 1924, 1955, 1958, 1961, 1964, 1967-68, 1978, 1988-89.[67]

The area to the west of the western margins of the Yamuna river is not free from floods either. Among the main causes of floods, uncontrolled supply of rain water brought by various seasonal streams, called as '**Nai Nallah**', coming from districts of Panipat and Karnal in the north, the flat terrain tendency of rising water-level, loam soil checking rate of percolation, the poor carrying capacity of the existing drains for their shallow nature and poor maintenance and the high rainfall upstream in the Siwalik hills, can be enumerated.

The people of this area have experienced notorious floods in the years of 1885-86, 1892-93, 1894-95.[68] Similarly, thousands of hectares of crop worth crores of rupees were affected by floods in 1924-25, 1933, 1960, 1961, 1963, 1964, 1967, 1981 and 1982, but the flood in 1933, was unprecedented causing unmeasurable hardships to man and animals. Likewise in the years of 1960-61 and 1963, floods ravaged badly villages, such as Baroda, Jagsi, Bhanderi, Banwasa, Chhapra, Bhawar, Kathura, Wazirpur, Gangesar, Bichpari, Khandrai, Dhanana, Chidi, Sewana, Butana and Ahulana.[69] This has been repeated in 1995, with unexpected damage to life and property. To enhance the efficacy of the existing anti-flood methods some other techniques such as drilling of deep bore tube-wells, construction of reservoirs and broadening, deepening and regular clearing of the existing drains can prove extremely useful to mitigate, if not rule out a flood menace to the life and property.

Famines and Droughts

If the actual rainfall happens to be below normal by more than twice the mean deviation, then that year can be termed a year of drought.[70] The areas where famines usually visit are those with irregular and inconsistent rainfall as is the case with the area under discussion. Such areas are more severely exposed to this danger if protective irrigation on an adequate scale is not provided. Dry spells in such areas could cause acute shortage of water and thus, famine knocks the door.

The record of famine is a sad chapter in the history of this district. With the insufficient means of irrigation, poor quality of subsoil water, especially in the western section, notoriously precarious rainfall and low technological expertize it was, but natural to experience famine conditions often. Each of these famines was given a specific name after the year of its occurrence, which is considered like an epoch in the history of the countryside. Famines and epidemics were two major scourges in the lives of the villages in medieval times.[71] Such conditions were experienced in the following years—1753-54, 1782-83, 1802-03, 1812-13, 1817-18, 1833-34, 1837-38, 1860-61, 1868-69,1877-78, 1886-87, 1895-96, 1896-97, 1899-1900, 1905-06, 1907-10, 1913-14, 1918-19, 1928-30, 1938.[72]

The foregoing detailed list of famines suggest that the area was exposed to such natural hazards because the temperature runs usually very high, rainfall is very unpredictably irregular and sources of irrigation including canals, wells and others were very nominal by the turn of the 19th century. And subsequently also these remained highly restricted in frequency and simple in nature till the forties of the current century in the absence of fast technological advancement. Therefore, under the given circumstances the frequency of natural calamities was not an unusual phenomenon. The recent ongoing drought conditions are causing serious concern despite all the modern technological developments. In case these conditions persist for some more years to come the water-level will go either beyond the economic range or completely out of reach. And areas with limited and shallow underground water reservoirs, comprising only to the upper part, will also go either dry or turn saline after some time converting the larger part of the district in a desert. Thus, the consequences will be very disappointing in the year to come, if corrective measures to arrange sufficient amount of canal water are not adopted in time. Such adverse impacts are already being felt in some parts of this zone.

Irrigation Network

The irrigation system of this area mainly depends on canals, tube-wells and rarely on wells now. The district is provided with a well-knit network of canals[73], but the frequency and intensity of irrigation is not sufficient for want of regular water supply to produce two to three crops annually, so essential to produce maximum foodgrains. About 232200 hectare of land is irrigated in this district. Of that 72700 hectare area is irrigated by canals alone. The place of tube-well irrigation is second with 47000 hectare of land. Wells used to cover only 63,000 hectare area, which is shifting to pump-sets and tube-wells as well irrigation is becoming out of fashion. About 49500 hectare of land is irrigated by other sources. Thus, it is amply clear that the largest area is receiving water from canals. These canals cover 55.58 per cent of the total irrigated area.[74] The canal network, of course, has a long history and, yet, has to go still a long way.

History of Canals

The Western Yamuna Canal was dug by Firuz-Shah for the first time to conduct water to royal gardens at Hansi and Hisar. Subsequently it was further enlarged and improved in Akbar's times to bring perennial supplies from the Yamuna and the Somb into the Chautang and on to the Hansi and Hisar. Later in the reign of Shahjahan it was further improved by Ali Mardan Khan in AD 1643, with the object to diverting the Yamuna water to Delhi. The river supply was tamed to 22 kilometre below the present headwork of the canal and the water was led along the water drainage line through Panipat and Sonipat and finally to Delhi.[75] Thus, for the first time canal water traversed through the district in AD 1643, only and this must constitute the beginning of canal system in this area.

With the fall of the Mughal empire, the canal, by then known as 'Shah Nahar', gradually silted up. Fortunately in 1821, a small instalment of the Yamuna water was diverted into Delhi canal adhering to the original alignment, which turned out unsatisfactory. After a lapse of almost 25 years in 1846, it was realized that the concentrated irrigation and defective drainage contributed to the rapid deterioration of the soil. Keeping in view these developments the canal was realigned and thus, assumed the present form. The present Delhi Branch is 847 kilometre in length and enters this district at Dhindhar in Panipat and leaving it between Halalpur and Jhinjhauli villages.[76] This canal forms the main hub for canal system of the district and supplies water for irrigation and drinking.

Minerals

Generally the district is devoid of minerals and for that matter cannot be rated very rich. 'Salt-petre'-potassium-nitrate and foundry sand are available in Gohana and Sonipat tahsils respectively. The former appears as efflorescence on the surface during the summer. Its encrustations are extensively worked around Gohana and Ganaur. It is used for manufacture of gunpowder, in crackers, in match industry and as fertilizer. Some contractors are exploiting it intensively. Notably a large number of ancient settlements are being spoiled by salt-petre making activities and thus, the district and the state are being robbed of their precious ancient cultural treasures, which cannot be recovered, if once lost, by any endeavour to reconstruct the past patterns of human activities in this area. The latter (foundry sand) occurs at Sonipat at a distance of about three kilometre from Sonipat in the east along the National Highway No. 1 (Shershah Suri Marg) leading to Karnal from Delhi. It is high grade silica sand left behind by the receding Yamuna waters. The chemical analysis gives about 79 per cent silica and the ignition temperature is 129°C.[77]

Flora

The vegetal cover at present is really very thin, but composed of a variety of trees, plants, shrubs and grasses generally found in different micro-environment nitches, such as dry land, sandy soils, low-lying areas, alkaline soils, waterlogged pockets and *Khadar* belt. The flora, thus, demonstrates a synthesis of various characteristic features of these zones. Let us now have a brief look on the specific plants of the respective micro-pockets in the following discussion.

To begin with we may mention some of the characteristic plant species of the sandy pockets. Among them Dholimundi or Dholphuli or Kamhera (*Aervapseuda tomentosa*), Jawansa (*Alhagi camelorum*), Rattanjot (*Arnekia hispidissima*), Jhadberi (*Zizyphus mumularia*), Ber (*Zizyphus oenoplia*), Kasani (*Chapparis aphylla*), Gokhru (*Tribulus terrestris*), Punarnva (*Boerhavia diffusa*), Beri (*Zizyphus jujuba*) and Burat (*Xanthium strumarium*) are very noteworthy. Jhadberi leaves provides a good grazing and dry fodder for animals in the off seasons. The leaves are generally stored for this purpose. Burat gives poor grains for human beings as well as offer fodder for cattle. It has the quality to spread rapidly and to become troublesome on account of its fruits with hooked bristles. Other plants found are Aak (*Calotropis procera*), Kala Bhangra (*Croton bon plan dianum*), Satyanasi (*Argemone mexicana*), Rassa (*Crucus arvensis*). These are wasteland plants and are found in drier conditions.[78]

Among the sandy area plants mention may be made of the following varieties like Kans (*Saccharum spontanium*)—a very troublesome weed, but when young easily grazed and chopped and stored for dry fodder in bad times; Jhund (*Anatherum muricatum*)—generally found in ponds and depressions is very valuable and useful for thatching and brooms; Makrah or Diela (quaere)—usually give a very good grazing during rainy season and winters; Sanwak (*Panicum colonum*)—useful for grazing in the monsoon, provide a good harvest to store and later used as dry fodder in winter, bears small bajra like grains used for eating in hard times; Kaundra (*Digera arvenesis*)—very useful for vegetable and grazing and Bathu (*Chenopodium album*) is very harmful for crops, but equally useful for vegetable and grazing in the winter season.[79]

In this connection some more plants of the rainy season such as Hulhul (*Cleome viscosa*), Kankawwa (*Commelina*), Kagaroti (*Corchorous*), Kulfa (*Pontulaca*), Santhi (*Trianthema*), Bhankhadi (*Tribulus alutus*), Kasni (*Triumphetta tomentosa*), etc., may be listed, which provide a green cover to the entire landscape so pleasing to the eyes. Similarly some other plants of muddy and marshy lands are also noticed, especially along the Yamuna. These are *Riccia*, Horse-tail and tiny Buttercup. Likewise Singhara (*Trapa bispinosa*) also belongs to aquatic plant and fruit group—very tasty and sweet before ripening. The ripe fruit becomes still more delicious after boiling. It is cultivated in and around low-lying ponds, depressions and areas during rainy season and produces fruits in November and December. There are some more plants known as crop weeds like Piyazi (*Asphodelus tenuifolius*), Dhudhi (*Ehrita*) and Nunka (*Phyllanthus fraternus*). All these plants, of course, offer a good grazing.

The main trees of the district are Pilu (*Salvadora oleoides*), which is now almost disappeared, Jand (*Prosopis*)—found very often in the Bhangar area, Kair (*Apparis decidua*), Heins (*Capparis *sepiaria*) and Tamarisk—locally called as 'Faransa' (*Tamarix articulata*). These trees grow rapidly after trimming without any water and are very commonly found in drier parts of the district. Shisham (*Dalbergia sissoo*) is a common tree found everywhere and well-known for its timber. Babul or Kikar-desi (*Acacia nilotica*) is a common tree of this area and invariably noticed in every nook and corner of this district. It suits most to the prevailing climatic, hydraulic and soil conditions for its tremendous salinity and drought resistance. It provides a good fuel, toothbrush and certain medicines to the people in the countryside. It also offers a good grazing for goats in normal times and in abnormal or hard times its green leaves are used as fodder. Similarly Neem (*Azadirchta indica*) flourish in these given conditions and found almost everywhere. It not only provides shade to man and animal, but also used as toothbrush and for medicinal purposes. Besides, there are also found some more trees, such as Siris (*Albizzia lebbek*), Dhak (*Butia monosperma*) and Banyan (*Ficus bengalensis*). Pipal (*Ficus religiosa*) is very popular for religious sanctity and sweet fruits.[80]

In addition to these there are some fruit trees also. Among them the most popular are Jamun (*Suzygium cumini*), Mango (*Mangifera indica*) and Mulberry (*Morus alfa*), Gullar or Udumbra (*Ficus cunia*) and Kaindu (*Diopyros cordifolia*) are also visible here and there. Some more trees of general character are also met with, which include Barna (*Creteva nurvalla*), Sohanjana (*Moringa oleifera*), Kaim (*Mitrogyna*), etc. These are not witnessed in fairly large number. There are some large shrubs, which associate themselves in close clumps, offer shelter and support climbers. The common climbers are Kanduri (*Occinea cordifolia*) and Aaksan (*Maerua arenaris*). Nagphani (*Opuntia dillenii*) is the common prickly cactus which may be seen in wastelands and along hedges.[81] Bansa (*Adhatoda vasica*) and Sambhalu (*Vitex negund*) are usually used as roofing material for mud houses. The former is also used as fuel and a protective mechanism along hedges against stray animals and soil erosion.[82]

Some grasses also do occur and a few of them like Dub (*Cyndon dactylon*)—a very common and good fodder grass for drought periods, Munja (*Sacharum spontaneum*)—large in size, provides fibre for making ropes, supplementary material for houses, screens, walls and thatching. Sometimes used for making principal structure either in absence of suitable technical hands and expertise or when the required material like good mud, brick and stone are lacking. The houses constructed of this grass are cool during the summer and warm in the winter. It is generally used by economically backward sections of society these days. Khaskhas (*Erograstis*) and Dab (*Erograstis cynosuroides*) can also be cited.[83]

The above mentioned flora differ in character and properties because some of them have wide range of tolerance for many factors. Such trees and plants find wide range of distribution over vast pieces of land. Some of the plants have a wide range of variant tolerance for one factor to another. Such plants cannot find wide circulation over the regions rather these can be seen flourishing in specific localities. It has also been pointed out in the foregoing discussion that the flora do reflect such influence in terms of quantity rather than quality since the number of various above mentioned plants underwent a negative change over the years. But the quality, however, remains more or less the same. And therefore, the district, which previously used to abound with thick cover of vegetation (locally known as 'Bir'), is gradually reduced to a bare 0.99 per cent forest cover of the total land. And the land is increasingly converted into agricultural fields. Such a growing decrease is dangerous to future agriculture, animal husbandry and the human life itself because the soil without a reasonable vegetal cover losses its capacity to hold nutrients or maintain its structure. The removal of vegetal cover causes the loss of upper soil, the humus—so rich in organic substances. It is virtually an irreparable loss and takes nature about hundred years to build an inch of it.[84]

Fauna

A large number of birds and animals usually come across in this district. These are antelope (blackbuck), *hiran* and gazelle (*chinkara*), which are occasionally visible in areas of Kakana, Sargthal, Gohana and Baroda in Gohana tahsil and Dobeta, Murthal, etc., in Sonipat tahsil. *Hiran* is still protected in some villages around Bhainswal, but by the villagers themselves. Hunters are not allowed to enter the revenue limits of these villages. Fox, jackal, wildcat and squirrel are commonly met with. Wolf, hyena and wildboar are now almost non-existent. *Nilgai* (blue bull) is gradually reappearing menacingly. Antelope is almost on the verge of extinction. Monkeys are seen, but not so frequently so in earlier times.[85]

So far as the birds are concerned we may enumerate ducks, geese, snipes, cranes and cocks found generally around the lakes of Kheri Sargthal, Rana Kheri, Bhainswan, Mundlana,Bali and Rithal in Gohana tahsil and Sersa, Murthal, Barwasani, Khubru and Dobeta in Sonipat tahsil. Partridges, pigeons and peefowls are visible over the entire district. Sand grouse is found in open, barren plains, stubble fields and land away from water around Kohla and Banwasa in Gohana tahsil. The other common birds include sparrow, crow, kite, vulture, babbler, parrot, kingfisher, hornbill, bulbul, cuckoo, baya, dove and woodpecker.

Among the reptiles crocodiles are not as common as snakes in this area. Even the poisonous varieties of snakes, such as cobra and krait are also occasionally found. Besides lizards are very commonly visible.

Regarding pack animals we may record the ass, horse, camel and ox. The last is very popular for ploughing fields. This is of a special Haryanvi breed, which is so popular in this area. For milk Haryanvi 'Murrah' breed buffalo is very famous. Cows are equally good and they play a significant role in socio-economic life of the people.[86]

Present Settlements : Number, Size and Distribution System

The district harbours 348 villages and four towns of Gohana, Ganaur, Kharkhauda and Sonipat. These villages are shared by two tahsils of Sonipat and Gohana. The number of villages in these two tahsils are 254 and 94 respectively.[87] This shows that the number of villages in the tahsil of Sonipat is more than double the strength of the Gohana tahsil. The villages in the tahsil of Gohana are sparse in comparison to the tahsil of Sonipat. But, interestingly on an average the villages in Gohana tahsil are not only large, but also evenly spaced. And as far as the locational analysis of these villages is concerned there is hardly any imbalance unlike Sonipat tahsil where villages are usually of small to moderate size and closely spaced. This tendency is quite prominent in the eastern half of the tahsil, i.e., the new alluvium, the modern flood plain of the Yamuna river, which is locally known as the 'Khadar' or the low-lying area. This feature is of recent origin and seems to have assumed significant proportions in only recent times, especially after the introduction of modern agricultural tools and techniques with the beginning of green revolution and consequent population explosion. This in combination with the prevailing favourable natural condition of abundant fresh water and alluvial fertile soil stimulated this process. The harmonious relation between these forces produced this system. But the very application of the new technology could not effect the same change in the Gohana tahsil mainly in the absence of similar natural conditions there. This shows that the two variables—natural environment and technology are effective only, when these are in harmony. The level of their effectiveness is indicative of the degree of harmony. Since the degree of this harmony was on the lower side in the Gohana tahsil, the degree of change and transformation in terms of number, size and frequency of settlements was also restricted unlike the tahsil of Sonipat. This is further elaborated by the fact that the western half of the tahsil of Sonipat is also a part of the old alluvium, where the natural features are not dissimilar to those of the tahsil of Gohana, since it merely forms the easternmost extension of the old alluvium, yet, the pace of development is just like that of the new alluvium, the Khadar simply because of the fair availability of canal water for agriculture. In comparison to the tahsil of Gohana this part of the land not only received canal water first, but also in greater amount for being better provided with canal network.

With regard to the locational system one may spontaneously notice that on the whole the villages are located on elevated pieces of land for safety, but within reach of some natural depressions, which at some stage have been transformed into village ponds. All the villages used to have more than one ponds in their close vicinity and several other in their revenue limits. There are also village wells. Some of them were sunk prior to the beginning of the green revolution and others only during that period. The former are situated near the ponds and depressions, while the latter are located along canal water-channels, especially in the dry belt. The reason for such an approach in both the cases was the availability of fresh subsoil water in upper stratas. And the purpose was to tap that natural resource to serve their requirements of water. The availability of water also affected the pattern of expansion of villages along with other factors as the villages are found enlarged along the line of water. Another very interesting feature of modern villages is the trend of agglomeration and branching off. In the Bhangar belt the practice of congregation is quite strong and usually people desist from branching off from the parent settlements until it becomes irresistible. Contrary to this in the new alluvium—the Khadar belt the temptation for budding off from the parent village is quite strong, which hampers considerably the growth pattern in terms of size of the villages. As a result the villages are relatively small in size and are located in close proximity with each other. The size of the villages in context of number of inhabitants in both the contrasting geographical zones points to the varying carrying capacity of the environment.

Background of Archaeological Studies

The history of archaeological studies travels back to the second half of the 19th century, i.e., 1871, exactly after a decade of the establishment of Archaeological Survey of India, when none other than Cunningham, the then Director-General, took initiative to explore the archaeological potential of northern India.[88] This was to ascertain the identity of historical places on the basis of literary narratives. In that process he identified the Sonaprastha of Mahabharata with the modern town of Sonipat after which the district under discussion was named. He also discovered a hoard of 1200 Indo-Greek hemi-drachams from the mound of Sonipat for the first time and thus, threw welcome light on historical background of Sonipat. This exercise, though limited in scope so far as the district and its settlement studies are concerned, was of enormous significance for the purpose of future archaeological studies. On the one hand it fixed the historicity of the area under review and on the other underlined the archaeological potential for prospective archaeological studies. However, there remained a complete lull for about a quarter century, until B.B. Lal began to explore the area to establish the identity of some of the historical places as mentioned in the epics. During that course of extensive survey he also visited the town of Sonipat and collected a few sherds of NBPW from here. These sherds confirmed the existence of the settlement of Sonipat in early historical period.[89] In the early sixties Dikshit intended to explore this belt on a somewhat largescale initially, though he kept himself confined to a particular belt along the railway track from Delhi to Karnal. This exercise increased the number of known archaeological sites by putting on record very few, but important settlements representing PGW and NBPW levels from this area.[90]

In the early sixties of this century Fleet reported the famous copperplate inscription of Harshavardhana, the king of Thanesar from this place (Sonipat). This refers to horse trade activities alone, yet, its historical importance is great.[91] Another inscription in Persian of Mamluk dated AH. 767 AD 1277, was reported from here by The Archaeological Survey of India in 1973.[92] This was issued by Ghiyas-ud-din Balban where name of *Muqta* of Sonipat is mentioned as QirBek. Similarly, Dikshit reported the site of Kheri Gujjar representing medieval period in 1968.[93] Undoubtedly all these activities are limited, isolated and disconnected to assume shape of any concerted attempt addressed to any specific problem. However, in the late sixties a significant directional shift took place with the conservation of Khawaja Khijer's tomb at Sonipat.[94] Such an approach was long overdue and still badly needed not to protect and conserve only medieval monuments, but also ancient settlements for their historical, social and archaeological significance. And in absence of any such effective approach the remnants of the past are open to ruthless destruction by natural and human agencies alike. Had it been taken care of at least in terms of proper recording, if not preservation, a vast repertoire of ancient knowledge would have been saved from being partial or complete loss for ever. This could have gone a long way to reconstruct past activities of people in this belt and the area at large.

Subsequently, Bhan undertook an extensive survey in order to establish cultural chronology of the region and to put on record as many settlements as possible.[95] In the same way Silak Ram set out to explore the area for his Ph.D. Thesis on '*The Archaeology of Rohtak and Hisar Districts of Haryana*'.[96] These attempts were part of an interesting innovation in the direction to explore more and more settlements. Thus, some parts of the district under review, as it formed part of district Rohtak earlier, received fresh attention. As a result not only a good number of settlements was placed on record, but also the cultural sequence was established. These attempts amount to the status of pioneering work. Nevertheless the settlement data were not viewed in the light of geographical, ecological, technological and ethnographic approaches. However, these works enjoy the fundamental status of being crucial to any future attempt. This was followed by reporting of two inscriptions of medieval period from the town of Sonipat. One refers to construction of a Tomb of Imam Nasir in AD 1271 and the second is related with a square pillar erected in AD 1479, near the grave of Imam Nasir.[97] Again there is a reference to a Mausoleum of Nasir erected by Khizar Khan and a Kos-Minar of Mughal period at a short distance from Sonipat. The Kos-Minars are indicative of a past highway to the east of Sonipat. This was followed by a novel approach to study the relationship between the town of Sonipat and its close environs with the help of archaeological data.[98] In that some more settlements were added to the galaxy of ancient sites of this zone.

Keeping in view the foregoing brief survey of the various research activities what is quite obvious is the outstanding need for an intensive district-wise combing of the region based on village to village exploration approach to put on record the whole range of information on past activities, before it is lost for ever under pressure of multifarious activities of modern development. There is, therefore, a strong requirement to undertake an intensive systematic survey of the district. And thus, the present exercise is in that direction to explore past patterns of sustained human existence and supporting subsistence activities with the help of settlement data on the past. This exercise is obviously primarily based on surface collection and its examination to understand past patterns of dispersal of production activities over the landscape. The prospective analysis would provide a perspective from, which the sequence and spread of settlements in the ancient past could be perceived.

NOTES AND REFERENCES

1. *District Sonipat Census Handbook—Part A and B*, 1981, p. 5; *Ground Water Potential of Sonipat District —Haryana State—* North-Western Circle Region, Chandigarh, 1983, p. 3.

2. *Ibid.*
3. *Ibid.*
4. *Ibid.*, pp. 6-10.
5. *Ibid.*
6. *Ibid.*
7. *Haryana District Gazetteers—Rohtak*, Chandigarh, 1970, p. 332.
8. Athavale, V.B., 'The Movements of the Pandavas', *ABORI*, xxix, 1948, pp. 85-95.
9. Agrawala V.S., *India as Known to Panini*, The University Press, Lucknow, 1953, p. 549.
10. Lal, B.B., 'Excavation at Hastinapur and Other Explorations in the Upper-Ganga and Sutlej Basins', *AI, 10 and 11*, New Delhi, 1954-55, pp. 5-151; Singh, Surender, *Archaeology of Sonipat*, M.Phil. Dissertation (unpublished), Kurukshetra University, Kurukshetra, 1984, p. 11.
11. Cunningham, A., *NC,* 1872, p. 161, Idem. *Coins of Ancient India* (reprint), Indological Book House, Varanasi, 1963, p. 118.
12. Phogat, S.R., *Archaeology of Rohtak and Hisar District of Haryana*, Ph.D., Thesis (unpublished), Kurukshetra University, Kurukshetra, 1972, p. 262.
13. Fleet, J.F., 'Sonipat (Sonepat) Copper Seal of Harshavardhana', *CII*, III, 1963, pp. 231-33.
14. Lal, *op. cit.*, p. 19; Singh, *op. cit.*, pp. 1-3.
15. Buhler, G. 'The Pehowa Inscription from the Temple of Garibnath', *EI*, 1, 1970, pp. 184-98, The date of this inscription is Harsh Era 276 (AD 882) and refers to horse trade at Pehowa. According to another inscription dated AD 1163/1164, on the Ashokan Pillar from Topara, Ambala district, Haryana, now in Kotla Firozshah, the area between the Vindhyas and Himalayas was captured by the Vigraharaj - IV (Vishaldeo/Bisaldeo C. 1153-64), of the Chahaman/Chauhan dynasty of Shakambhari, *IA*, xix, 1890, pp. 215-19.
16. Eggeling, J. 'Delhi Museum Inscription' *EI,* 1, 1907, p. 93, Yadava, K.C., *Haryana ka Itihas* (in Hindi), 2nd. ed., 1981, p. 20; Buddha Prakash 'The Earliest Reference of Haryana', *Haryana Studies in History and Culture*, Kurukshetra University, Kurukshetra, 1968, pp. 42-46. According to him the earliest reference to Haryana in the form of 'Haryanaka' occurred in the Ganesgarh Plate (Saurastra, Gujarat) of the 6th Century AD and was given by Abhiras hailing from Haryana.

17. *EI*, 1, 1892, pp. 93-95; Handa, Devender 'Some Important Towns of Haryana', *JHS*, II, 1-2, 1970, pp. 4-21; *JHS*, III, 1, 1971, pp. 8-16, 2, 1971, pp. 1-8; Habib, Mohammad, *Sultan Mahmud of Ghazni*, Delhi, p. 128.
18. Horn, Paul, 'Muhammadan Inscriptions from the Subah of Delhi', *EI*, 11, 1970, pp. 130-59; *Haryana Gazetteers*, 1970, *op. cit.*, pp. 20-21.
19. *IAR*, New Delhi, 1973-74, p. 43.
20. Phogat, S.R., *Inscriptions of Haryana*, Vishal Publications, 1978, p. 121.
21. *Haryana Publications*, p. 32; *District Census*, p. 9.
22. Phogat, 1978, p. 71.
23. *Ibid.*
24. Fanshawe, H.C., *Assessment Report of the Main Part of the Rohtak District*—1873-79, 1879, pp. 1-35; *Haryana Gazetteers*—Rohtak, 1976, p. 3.
25. Canning, F.C., *Assessment Report of the Sonipat Tahsil of the Delhi District*, 1881, pp. 1-27; *District Census*, pp. 9-11.
26. *Delhi District Gazetteers*, Delhi, 1976, pp. 2-3.
27. *District Census*, pp. 9-11.
28. *Haryana Gazetteers*, 1970, pp. 326-27.
29. *Ibid.*
30. *Ibid.*
31. *Ibid.*
32. *Punjab District Gazetteers*, Delhi, 1910, p. 1.
33. *Haryana Gazetteers*, 1970, pp. 326-27.
34. *District Census*, pp. 9-11.
35. *Ground Water Sonipat*, pp. 6-7.
36. *Ibid.*
37. Joseph, E., *Third Regular Settlement Report of Rohtak 1905-10*, 1910, pp. 1-2.
38. *Ibid.*, *Haryana Gazetteers*, 1970 *op. cit.*, pp. 4-5; Singh, J., *An Agricultural Geography of Haryana*, Kurukshetra, 1976, p. 36; *Delhi Gazetteers*, 1976, *op.cit.*, pp. 2-5; Raychaudhury, et.al. *Soils of India*, ICAR, New Delhi, 1963, pp. 435-48.
39. *Haryana Gazetteers*, 1952, *op. cit.*, pp. 9-12; Kuhad, M.S., *Study of Morphological and Soils of Block Sonipat (Rohtak-Haryana)*

and their Classification. M.Sc., Dissertation (unpublished), Hisar Agriculture University, Hisar; Singh, J., *Study of Morphological and Physiological Characteristics of Soils of Block Ganaur (Sonipat-Haryana) and their Classfication*. Ph.D., Thesis, (unpublished), Hisar Agriculture University, Hisar, 1971, pp. 54-63.

40. Kuhad, 1970, *op. cit.*; Kaushik, R.N., *Comparative Study of Normal and Salt Affected Soils of Ganaur*, M.Sc., Dissertation (unpublished), Hisar Agriculture University, 1972, pp. 631-42; Sindhu, G.S., et al., 'Soils of Haryana', in Raychaudhury, S.P. et.al. eds. *Soils of India*, ICAR, New Delhi, 1963, pp. 106-11; Duggal, S.L. *Soil Geographical Zones of Haryana*, Hisar Agriculture University, Hisar, 1970, pp. 1-4.
41. *Ibid.*
42. *Haryana Gazetteers*—1970, *op. cit.*, p. 5.
43. Beadon, H.C. and I.A. Major, *Third Regular Settlement Report of Delhi District, 1906-10*, 1910, pp. 1-2.
44. *Delhi District Census Handbook*, Part C, Delhi, 1971, p. 1.
45. *Haryana Gazetteers*—1970, *op. cit.*
46. Joseph, *op.cit.*, pp. 1-2.
47. Singh, R.L., *India: A Regional Geography*, 1970, p. 92; Duggal, S.L., *Soil Geography of Haryana*, Hisar Agriculture University, Hisar, 1975, pp. 3-5.
48. *Haryana Gazetteers*, 1970, *op. cit.*
49. *Haryana Gazetteers*, 1970, *op. cit.*, pp. 70-71.
50. Personal Communication with Satvir Singh Phogat, Incharge of the Soil and Water Testing Laboratory, Sonipat—Government of Haryana.
51. *Punjab Gazetteers*, 1910, *op. cit.*, pp. 100-101.
52. *Ground Water Sonipat*, *op. cit.*
53. *Haryana Gazetteers*, 1970, *op. cit.*, p. 66.
54. *Ground Water Sonipat*, *op. cit.*, pp. 7-8.
55. Singh, 1970, *op. cit.*
56. *District Census*, *op. cit.*, pp. 5-7.
57. Personal observation during the course of exploration of the district.
58. *Delhi Gazetteers*, 1976, *op. cit.*, p. 9.
59. *Haryana Gazetteers*, 1970, *op. cit.*, p. 99.
60. *Ibid*, p. 61.

61. *Haryana Gazetteers*, 1970, *op. cit.*, pp. 17-18; *District Census*, *op. cit.*, p. 9.
62. *Ibid.*
63. *Ibid.*, pp. 18-19.
64. *Ibid.*
65. *Ibid.*, pp. 19-20.
66. *Ibid.*, p. 62; *Agriculture: Report on the Operations of the Department of Agriculture*, Punjab, Official Annual Publication, 1905-47, issued by the Director of Agriculture.
67. *Punjab Gazetteers*, *op. cit.*, pp. 21-22; *Agriculture Report*, 1905-47, *op.cit.*
68. *Haryana Gazetteers,* 1970, *op. cit.*, p. 98.
69. *Ibid.*
70. Chatterjee, S.P. and T.B. Lahiri, 'Man and Environment—A Trend Report', *A Survey of Research in Geography*,—ICSSR, New Delhi, Popular Prakashan, Bombay, 1972, p. 236.
71. Satish Chandra 'Standard of Living in Mughal India', Raychaudhuri, T.P. and Irfan Habib eds. *The Cambridge Economic History of India*, I, reprint, 1993, pp. 458-71.
72. Joseph, *op. cit.*; *Agriculture Report 1905-47*, *op. cit.*
73. *District Census*, *op. cit.*; Bhardwaj, K., *Irrigation in India: Alternative Perspective*, ICSSR, New Delhi, 1990, pp. 5-15; *Agriculture Report 1905-47*, *op.cit.*
74. *Ibid.*
75. *Haryana Gazetteers*, 1970, *op. cit.*, p. 63; 1976, *op. cit.*, p. 121.
76. *Haryana Gazetteers*, 1970, *op. cit.*, pp. 63-64.
77. *GSI Publication*, 1976.
78. *Ibid.*
79. *Haryana Gazetteers*, 1970, *op. cit.*, pp. 13-18.
80. *Ibid.*
81. *Ibid.*
82. *Ibid.*, pp. 10-15.
83. *Ibid.*
84. *Ibid.*
85. *Haryana Gazetteers*, 1970, *op. cit.*, pp. 14-16.
86. *District Census*, *op. cit.*, pp. 5-9.
87. *Ibid.*

88. Cunningham, A, *Archaeological Survey of India Reports* (four), 1862-63, 64-65, II, 1871, pp. 54-60, 212-30.
89. Lal, 1954-55, *op. cit.*
90. *IAR,* 1960-61, p. 65.
91. Fleet, *op. cit.*
92. *IAR*, 1973-74, p. 43.
93. *IAR*, 1968-69, p. 9.
94. *Ibid*, p. 95.
95. Bhan, S., *Mitathal Excavations and Other Explorations in the Sutlej-Yamuna Divide*, Kurukshetra University, Kurukshetra, 1975, pp. 121-26.
96. Phogat, 1972, *op. cit.*, pp. 32-69.
97. Phogat, 1978, *op. cit.*, pp. 70-71.
98. Singh, 1984, *op. cit.,* pp. 1-3.

2

EXPLORATION

Introduction: Aims and Objects

Collection of settlement data by way of exploration and its interpretation form the main subject matter of this chapter. In order to examine variations in settlement patterning at micro (local) and macro (regional) level through time and space representing various cultures, it is essential to have as many settlements as possible to understand the manner in which these communities were arranged over the landscape. The distribution of the communities may provide an index of their intimate relation with the then prevailing natural conditions. To recover the possible basis of their relations with the environment demands the following methods and techniques.

The exploration process of this type in itself is a constituent of a larger discipline–Archaeology–and needs a scientific approach not only to discover ancient sites, but also to unravel the past patterns of subsistence. And these objectives can be achieved in a better way by adopting the following strategies:

(a) First of all it is called for to obtain information regarding already explored settlements from primary and secondary sources. And these sites were plotted on the map of the district to have an idea about the cultural sequence and their respective distributional position. This is extremely essential not only to form a stepping stone for the coming exercise, but also to have an idea about the prospects of such an exercise.

(b) The next prerequisite to proceed with the work was to have related Topo-Sheets from the Geographical Survey of India. Since these sheets invariably contain almost all the minor details about the area ranging from topographical information to remnants of past habitations, they can safely be relied on. And keeping in view the requirement such sheets were collected in the very beginning.

(c) To be more precise the necessity was to procure a recent drainage map of the district showing existing drainage lines, which not only provides guidelines to comb the area along them, but also indicate the probable past drainage pattern. This may further help explain their bearings on human life, if any.

(d) Keeping in view the urgency of having firsthand information on the existing hydrological scenario, canal maps were also secured. Such maps could furnish information on possible sources of water and the process of transformation of hydrological regime over the years. This brings out the comparative possible picture of past and present water sources. This could further help trace the corresponding impact on life-styles of people through times.

(e) Having equipped oneself with these pieces of information one is supposed to be capable of undertaking exploration work. And the survey must be of intensive nature based on village to village exploration. It is a must to achieve basic objective of getting more and more information of past settlements for appropriate analysis. And it is desired that no settlement, whatsoever remains unnoticed. To serve this purpose one is supposed to take into consideration the already explored sites on the one hand and on the other to visit every village without fail to seek information about the ancient remains from the villagers, especially the seniors, farmers and cowherds. These people tread the surroundings through time and gain knowledge of past traditions in that process. Similarly the respective revenue officers of the village level could also lend a helping hand being intimately related with the village environs and hence, such, officers were also contacted, wherever it was possible.

Efforts were also made to glean some more information on soil types, nature of vegetation, patterns of land use and subsistence. All these factors are closely associated with human life and could help a great deal to understand the past subsistence patterns in this area, i.e. the human struggle against natural forces. The present exercise began in the last week of April 1985 and continued till the end of July 1987. Its first phase began in the last week of April 1985 and continued for 12 weeks till the 3rd week of July 1985, the arrival of monsoons. This period was deliberately opted because of its being the harvest season most of the fields are clear, which increases visibility and help identify ancient remains, soil types, water resources, vegetation, temperatures and human responses. During this period the area of Khadar, the Yamuna flood plain, lying in the eastern parts of the district, in between the Karnal railway track in the west and the Yamuna in the east, was combed. Immediately after the rains, I set out again to revisit the area and remained in the field for four weeks in the month of October 1985. This time my purpose was to reinvestigate the area on criss-cross pattern to make sure that no remains were left unnoticed. Besides I intended to collect more information on rainfall, drainage, water-level, crops, pastoral activities and the landscape to plug any possibility of creeping in a jaundiced view of the situation. My third stint, the smallest, for one week began in the first week of January 1986 and this time I visited a few villages along the Yamuna and west of the Grand Trunk Road, the Shershah Suri Marg No. 1, with a view to have a look on the landscape during the winters. Hence, the 17 weeks spent in the field proved very useful to aquaint myself with not only the past remains, but also the ecology, natural resources, subsistence patterns and the people of this belt. In other words this was to make out probable lines of human relations with the nature in the given operational ambience.

In the third week of April 1986, my second phase of exploration began, which continued till the second week of July 1986. This time the area of concentration was along the western margins of the Yamuna, the high ridge belt dividing the Bhangar and the Khadar and the area lying in between Sonipat and Gohana towns in the west. During this span of 14 weeks interesting information of varied nature on the aforesaid aspects was gathered. And this offered a contrast in certain respects than the previous zone, which further inspired me to go in for more details. To get further information another attempt was made from the third week of October 1986, to the end of November 1986, the period coinciding with the 'kharif' harvest and 'rabi' sowing season. In this process certain areas around Kharkhauda, Rithal and Khanpur villages were re-examined on criss-cross pattern. With the result settlements like Sisana-II and Khanpur-II were discovered. In addition to these more information on the climatic condition, drainage and subsistence system was obtained. To have a fresh look of the environment during winters two weeks were devoted in the month of February 1987, which was further enlightening. Thus, 21 weeks time was spent in this zone.

The area west of Gohana was explored in the summer of 1987. This time I undertook this work on 27th of April and continued till 30th of June 1987. The entire western pocket was thoroughly combed during this period. A large number of settlements was put on record in addition to information on natural drainage, soil types and their combinations, hydrological scenario, patterns of land use, sources of water supply and vegetal cover. The antiquity of human remains, their cultural sequence and the substantial number of settlements was in stark contrast to the extant natural resources. On my way back to home I ventured once again to investigate the western margin zone around Balagarhi and Harsana Kalan villages. Here I came to know about the Yamuna silt deposition about 100 feet deep with potable water with the help of a newly sunk boring. This is very useful information. With a view to adjudge afresh the dynamics of natural resources base during rainy seasons and to gather more artifacts exposed by rains, I spent the month of September in the field. As a result some more sites like Rindhana V and Banwasa were recorded along with a better understanding of the landscape and its carrying capacity. Later on a final corrective attempt was made in the month of December 1987, around Ghadwal-Rindhana villages. However, no new settlement was noticed. Thus, in this area I spent almost 16 weeks to get the required information.

It has already been pointed out that 54 weeks were spent in the field. During this period I traversed every nook and corner of the district under review on a two-wheeler. But, wherever, it was not possible to ply on the two-wheeler I had to walk down through the fields. Always it was not possible to follow roads, though every village was provided with them, as ancient sites were not necessarily fall on them. In such circumstances *kuccha* roads, canal and drain sides, railway tracks and natural watercourses were followed. More often than not I had to walk down, even across the fields, barren lands and occasionally through village forests with a bag on my back and a water bottle in my hand to brave off the scorching sun of May and June. However, the whole exercise was very refreshing.

During the course of exploration I used to fillup a proforma meant for archaeological explorations for each and every village and site wherein columns were provided for every minute details like the name, location, block, tahsil, district, state, culture/s, conditions, shape, size, height, axis and surroundings of a site. The surroundings of a site cover soil types, their contours and patterns of use, depth and quality of subsoil water, depression, watercourse and river with orientation, size and nature, vegetation with its nature, degree of density and uses, facilities of irrigation—well, tube-well, canal or any other source. Besides details of salient features of the extant ancient remains, their relative frequency on the site/s in any particular portion of the site to assess, though tentatively, their respective area of frequency and their time-scale. The details about a site would remain incomplete if one fails to record other ancient remains in the immediate vicinity, their distance, direction and nature. Rough contours of the ancient mounds are also called for to enrich the recording further, which I could not follow strictly because most of the sites were found to be converted into agricultural fields. Thus, I accomplished the work undertaken by me in 1987.

As a result of the above exercise 106 settlements representing Protohistoric and early historic periods were added to the archaeological map of the area in addition to several other sites belonging to the Kushanas, Guptas and medieval periods. Previously reported 39 sites of the Protohistoric and early historical periods were also revisited. These explored sites can be described in various ways, such as in alphabetical order, culture-wise, geographical zones-wise and along administrative units. But for our own convenience to study the data, especially in this chapter, a tahsil-wise-cum-alphabetical framework is preferred. And since, the district consists of mainly two tahsils, namely Gohana and Sonipat, it is, therefore, desired to dwell on accordingly in the following paragraphs. But, before we begin to describe the settlements it is imperative to elaborate a few terms and categories according to which the explored data are classified and addressed to in the ensuing discourse. These terms include early Harappan, mature Harappan, late Harappan, OCP, PGW, RW/NBPW, etc.

Here the term early Harappan (here after EH) is used for a phase and its cultural assemblage, which precedes and relates with the mature Harappan cultural assemblage. Its specific attributes have already been identified at the type sites of Kalibangan, Banawali, Rakhigarhi and Mitathal in the adjacent area in the west. However, in the area under review all the characteristic features identified at the type sites were not available during surface exploration. Nevertheless, wherever, some of the traits of this early phase were noticed, these were treated as early Harappan. The ceramic shapes of this phase are vases, bowls, basins, dishes and troughs. These are decorated with simple linear designs in black, white and chocolate colours over plain red surface. There are also incised decorations on the inner and outer surfaces in the form of lines, single or multiple, dots, nail and comb marks. Herc is a heavy reliance on ceramic wares and obviously so because pottery happens to be the major constituent of any cultural assemblage during the period of our study let alone the early Harappan phase. Hence, such remains are grouped under the title early Harappan.

In the context of the mature Harappan (here after H) phase the choice of identifying characteristic markers is greater than the previous one. Here the pottery is bichrome with black painting designs over red polished surface. The main designs are pipal leaf, fishscale, peacock, triangles and rhombs. Important shapes include dish-on-stand, S-shaped Jar, goblet and beaker. Besides parallel sided chert blades and cubical weights, a variety of carnelian beads and art of writing are among the salient features. Wherever some of the features were marked we assign them with the mature Harappan phase.

Regarding the late Harappan (here after LH) phase things are more obvious than any other phase. Here the important point is that the mature Harappan traits degenerate substantially. The ceramic ware is still bichrome with black-on-red combination, but neither the shapes nor the painting designs nor the other cultural constituents exhibit the hallmark of the yester-years. The main shapes are dish-on-stand with drooping rim, simple vases, bowls and jars with splayed out rim. Painting designs are linear and repetitive over plain or red-washed surface. Precious stones are replaced with red sand stone and faience. Thus, remains with such features are taken as late Harappan.

The Ochre Coloured Pottery (OCP) is roughly contemporary to the late Harappan phase. Its circulation is strictly limited. It is thin in section, as compared to the late Harappan ceramic tradition, ochre coloured, not always well fired and therefore with powdery surface. It is invariably devoid of any surface treatment. Decoration is executed with the help of incision technique. The designs are consisting of linear patterns like lines, single or multiple, vertical or horizontal, slashes, notches in registers and flowing lines mainly confined to neck and shoulders. Such ceramic remains have been identified as Ochre Coloured Pottery.

In the succeeding period an entirely new ceramic tradition is encountered, which is known as the Painted Grey Ware (PGW). The pottery is thin in section, steel grey in colour, well fired and decorated with jet black colour. The painting designs are geometrical and natural. Main pottery shapes are comprising dishes, bowls and basins with incurved rims and miniature vessels. *Ghat*-shaped terracotta beads are specimen of this time. Such distinct features on any site help assign them to this group. Occasionally a few sherds of Black Slipped Ware have also been marked, which have been abbreviated as BSW in the succeeding discussion.

The next period is the early historical (C. 600- 200 BC) and during this period there are two types of ceramic wares the Red Ware (RW) and the Northern Black Polished Ware (NBPW). The former is profuse, whereas the latter is occasionally met with. However, the Red Ware we are dealing with is thus, contemporary to the Northern Black Polished Ware. This ware is wheel thrown treated usually with a red wash and occasionally with a slip, painting designs in black consisting of lines, bands, loops, etc., are confined to neck and shoulder zones. Incision and stamp designs are also found. Main pottery types include vases, jars, bowls, basins, dishes, lids, etc., in a good range of variants. Such a ceramic assemblage is treated as 'Red Ware'.

The NBPW is of usual type, but found in bits. Therefore, it becomes difficult to perceive their precise shapes. However, some shapes point to tneir being bowls and basins. Thus these two ceramic traditions of this period are addressed to simultaneously. A description of the explored sites is as follows.

Gohana Tahsil

This forms the western section of the district and the Bhangar dry belt. Basically soils of this area are aeolian in composition and appear level in general with a mild tendency to slope down in the south-east, which, in fact, is interrupted by intervening depressions and rising sandy patches of varying proportions. But if it is viewed on micro-level the area can be easily termed of undulating nature, which, in itself, is very significant to explain and establish levels of interaction of the past societies. Natural water sources either in the form of surface or subsoil do not seem fairly conducive to any large sedentary agricultural communities. And overall ecological conditions also do not appear supportive as well. In a way in the absence of modern amenities and advanced tools and techniques the whole landscape would assume a very hostile look. But the existence of a fairly large number of ancient settlements is enigmatic and calls for rational approach to analyse the problem, which is anyway to be taken up in the next chapters. However, the description of the explored settlements is as follows (map 4).

Description of the Explored Settlements

1. Ahulana – I: EH, H and LH 29° 07' N 76° 39' 04" E

The ancient site is located about 300 metre east of the village Ahulana. The entire site is under cultivation and therefore, it is not in good shape. Its cultural association is attested to basically by the presence of potsherds representing early Harappan, mature Harappan and late Harappan levels of the Indus Civilization. So far as the cultural deposition is concerned it is difficult to gauge it, since the site is in dilapidated shape. And regarding its expansion we are left with no option, but to make conjecture with the help of spread of potsherds over the area. The approximate size of the site is 990 square metre.

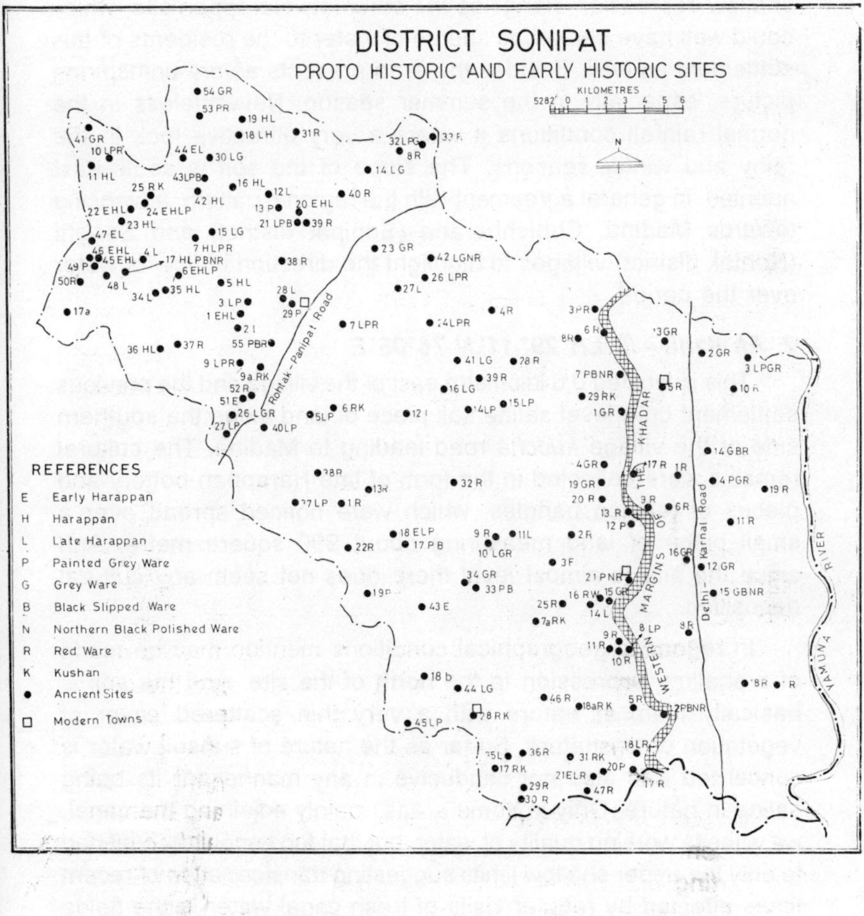

Map-4: Protohistoric and RW-NBPW Sites

The surroundings of this settlement do not tend to present any conducive environment. However, the ancient remains are located on gently rising aeolian soil without any source of water in the form of a depression in the close vicinity. But surely at a distance of about 0.5 kilometre in the south-east there exists a shallow depression alongside the other late Harappan site, which could well have served as source of water to the residents of this settlement as well. Besides the area presents a very uninspiring picture, especially in the summer season. Nevertheless in the normal rainfall conditions it wears a very attractive look in the rainy and winter seasons. The slope of the soil is south-east oriented, in general agreement with the regional pattern, advancing towards Madina, Chhichharana (Sonipat district) and Sanghi (Rohtak district) villages to highlight the direction of flow of water over the period.

2. Ahulana – II: LH 29° 11' N 76°05' E

This is located 0.6 kilometre east of the village and the previous settlement on a level saline soil piece of land along the southern side of the village *kuccha* road leading to Madina. The cultural remains were collected in the form of late Harappan pottery and pieces of faience bangles, which were noticed spread over a small piece of land measuring about 990 square metre. And since the site is almost level there does not seem any cultural deposition.

In regard to geographical conditions mention may be made of a shallow depression to the north of the site. And the soil is basically loam in nature with a very thin scattered cover of vegetation of dry nature. So far as the nature of subsoil water is concerned it is also not conducive in any manner for its being saline in nature. Only in some areas, mainly adjoining the canal, we witness working quality of water, but that too especially confined to only the upper shallow limits suggesting transformation of recent times effected by regular visits of fresh canal water in the fields for irrigation. So what formed the main stay in respect of drinking water for the people in the past is interesting point to be explored.

3. Ahulana – III: LH and PGW 29° 03' N 76° 52' E

This site is situated 2 kilometre north-east of the village Ahulana as the crowflies and just west of the Butana Minor running north-south. Locally it is called as 'Kakreda'. Only a small part of the mound is intact at present and the major portion seems to have been converted into agricultural field. Though earlier it used to cover an area of 24280 square metre and its height was 0.91 metre as reported by Silak Ram[1]. The cultural association is ascertained with the presence of late Harappan pottery and faience bangles on the one hand and PGW on the other.

Regarding the micro-level natural environment it is suggested by prevailing conditions that on the whole it is not appealing. Because frequent recurrence of saline soil patches and saline subsoil water quality are prominent. And the vegetational cover is also in agreement with them. A shallow and small depression is marked to the north of the mound. Therefore, it seems that such natural depressions might have served their water requirements in the past.

4. Anwali: RW and E. Med 28° 59' 07" N 76° 45' 01" E

The ancient remains are situated to the east of the village and near the pond. The remains are not in their natural position and amazingly some parts have been engulfed by the present day pond. Red pottery of NBPW period is found spread over the surface. But more interestingly some walls built with burnt bricks of early historical period were exposed during the annual clearing of the pond at a depth of 3 metre from the surrounding surface level and 0.8 metre below the bottom level of the pond. These walls were aligned with one another at right angles. These discoveries make one point very clear that is the location of the ancient habitation alongside the water containing depression, which was perhaps transformed into a tank keeping in view the increasing demand of water, the erratic rainfall pattern and saline nature of soil water. This highlights the change and continuity in patterns of human adaptability to his immediate environment.

5. Bali – I: LH and PGW 29° 12' N 76° 54' E

This stands 2 kilometre west of the village Bali and approached by a *kuccha* passage. The site is completely disturbed. Even though its cultural identity is borneout by the presence of late Harappan pieces of dish-on-stand and faience bangles and greater frequency of PGW. Although the 'saltpetre' making agencies have damaged the site to measure it out in an accurate manner, yet on the basis of circulation of cultural artifacts and their probable spread its approximate size comes to 2040 square metre with north-west to south-east axis.

The overall geographical scenario is not healthy. Because on the one hand salinity is rampant on the surface soil and the sub-soil water is also brackish on the other. The general nature, of course, of the soil is loam and the average vegetation is also of salt and drought resistant nature, fully in consonance with the existing conditions. However, there is a very shallow depression to the north-east of the site.

6. Bali – II: RW and Med 29° 03' 07" N 76° 44' 05" E

The ancient site has already been reported,is situated just east of the village near a pond. Its orientation is in the north-south directions. It was completely disturbed, which made it difficult to make out its size, but the spread of potsherds points to its being a big settlement in the past. And its big size was also recorded by Silak Ram. Its height at the time of Silak Ram's exploration was about 4.5 metre, whereas now it stands levelled[2]. Potsherds of early historical and medieval times are freely available to speak of its later cultural association.

7. Banwasa: LH 29° 08' 04" N 76° 34' E

The ancient site is situated to the south-east of the village and across the Gohana-Rindhana Road. Remains of late Harappan culture were found on an aeolian sandy mound covering a small area of about 990square metre without any sign of cultural deposition for being under cultivation. The nature of subsoil water is saline.

8. Baroda – I: H and LH 29° 08' N 76° 36' 09" E

The ancient mound is about one kilometre east of the village, east of a local drain and to the left hand side of the Gohana-Rindhana Road. The ancient remains in the form of Harappan and late Harappan pottery are superimposed upon gently rising aeolian sandy soil. The ancient remains are disturbed by agricultural activities as the mound is partly converted into agricultural fields. But, despite these disturbances remains of these two cultural strands are very prominent. If we go by the spread of potsherds its rough expansion comes to 3000 square metre. It is difficult to judge the thickness of cultural deposition for its disturbed nature. The longer axis seem to have been in north-west to south-east direction.

So far as the geographical condition is concerned we can simply put on record gently rising aeolian sandy soil around it, which is considered good for several purposes. The subsoil water is not of good quality and is deep-seated to about 20 to 25 feet. There seems to have been a depression to the west of the mound where the local drain runs through.

9. Baroda – II: EH, H, LH and PGW 29° 07' 03" N 76° 36' 07" E

The ancient mound is locally known as 'Gusain-Wala-Khera' and is located 1.5 kilometre away in the south-west direction from Baroda village, approached by a narrow *kuccha* village passage. The scatter of ancient remains speaks of its first ever occupation by early Harappan people followed by Harappan, late Harappan and PGW using people, though these remains are completely disturbed by human as well as by natural agencies. Hence, its precise orientation and expansion is of tentative nature. The approximate longer axis seem to have been in the north-south direction, if scatter of pottery is any indication. And similarly its size is estimated to about 4020 square metre. So far as the height of cultural deposition is concerned it does not seem to cross 0.6 metre mark.

The aeolian sandy rich soil around, deep-seated water-level of brackish nature, deep depression in the north and north-east, and poor vegetation cover point to the prevailing ecological conditions. Wells and tube-wells, confined to shallow aquifers, are successful only along the canal, which points to transformation of subsoil water only after introduction of the canal system.

10. Baroda – III: H, LH, PGW and RW 29° 09' 09" N 76° 35' 02" E

It was found located 2.5 kilometre west of the village Baroda and just east of the canal from Nuran Kheri to Madina. The ancient mound is known as 'Dab-Wala-Khera' locally. The longer axis runs east-west and covers an area of 4020 square metre. Since the entire site is under cultivation, there is hardly any cultural deposition, though the remains exist on an elevated aeolian sandy place. As per the cultural remains from here the site was first established in Harappan times and thereafter remained in occupation during late Harappan, PGW and RW levels. Protohistoric remains were found mainly from the western margins.

The geographical scenario is very interesting as composed of rising sand-dunes of varying proportion (aeolian origin) and low-lying pockets interspersed with level soils. A sharp depression exists on the north of the mound and comes from Nuran Kheri and Isapur Kheri via Kohla and Janata Butana College. Half a kilometre away in the west an aeolian sandy ridge, running north-south with low-lying areas to the west, is also quite significant. Subsoil water is workable supporting wells and tube-wells around the site, but this is mainly restricted to shallow aquifers. This is caused by frequent visit of canal water in this area, therefore this transformation of upper hydrological realm is of recent origin. Below the range of this transformed level and also away from canals the quality of soil water still continues of brackish nature. However, the geographical conditions are quite promising in all seasons.

11. Barota : LH-OCP, PGW and RW 29° 07' 20" N 76° 44' 35" E

It lies north-west of the village Barota and 200 metre north of Gohana-Sonipat Road on the rising aeolian sandy mound, which is now completely under cultivation and hence in a very bad shape. There is hardly any apparent sign of cultural deposition. The estimated area of the mound as per the spread of potsherds is 4335 square metre. Pottery remains of the first phase, i.e., the late Harappan-OCP are very exceptionally significant as such remains have been recorded from only a very few settlements. Still more interesting is the point that all such remains come from similar geographical locales—high sandy places overlooking low-lying and undulating soils. Potsherds of this culture do exhibit composition of light red of yellowish nature and grey colours alternatively on either side, which is, perhaps, promising to provide an insight into the evolutionary stages of full-bloomed PGW. Besides pottery types of PGW and RW were also picked up from the site especially from northern and north-eastern portions.

The information regarding the surrounding environment is provided by a combination of sandy, loam and clayey soils of shallow low-lying nature in the east and south, deep-seated water-table of saline quality and thin vegetation cover of drought resistant plants. This combination indicates existence of hard conditions and matching negotiation by the people in the past.

12. *Bhadauti Khas: RW and Med. 29° 14' N 76° 45' 02" E*

The ancient site is situated to the south-west of the village at a distance of about 2 kilometre on slightly rising aeolian sandy soils. It is completely under intensive cultivation and the spread of RW is very obvious. Its size is about 17030 square metre and lacks in cultural thickness for being disturbed.

The adjoining area consists of level loam and clayey low-lying soils with saline pockets in between. The water-table is of medium depth and the quality of water is not good for its being brackish. To the north of the site exists a shallow depression coming from district Karnal and which ultimately joins the big depression of Gohana. These days a small drain runs along the lines of this depression. The nature of vegetation is akin to the natural conditions. Thus, the overall emerging picture is not very conducive and that too especially in the summer season. And even these days the life of people is very hard.

13. *Bhainswan Khurd: LH, PGW and RW 29° 05' 50" N 76° 40' 50" E*

This mound is close to the north-eastern edge of the village Bhainswan, where government school is situated to the west of the railway track from Gohana to Rohtak. The mound, in fact, is partly covered by the school building and compound, but still a sizeable north-western portion is outside the school boundary. This uncovered portion is now under intensive agriculture. And the entire site is situated on a natural mound of sandy loam soil. Its cultural identity is established by occurrence of late Harappan ceramic wares and faience objects as well. Other pottery types of PGW and RW speak of its continuation in the subsequent periods. The spread of pottery over an area of 4020 square metre highlights its north-west to south-east extension in agreement with the general drainage lines.

The surrounding landscape is uneven, dotted with high and low-lying pockets consisting of aeolian sand and hard clayey black soils respectively. The subsoil water is of working nature with a depth of 15-20 feet below surface. And this is also in those areas, where canal water is used for irrigation for last many years. Anyway this is of recent origin. Beneath it the quality of water is once again brackish to saline.

14. Bhawar – I: LH, PGW and RW 29° 11' N 76° 30' E

This stands 1.5 kilometre north of village Bhawar. A tube-well with working quality of water is situated on the site. The entire mound is completely disturbed partly by farmers and partly by saltpetre making agencies, though its size and height are said to have been considerable as reported by Silak Ram[3]. The longer axis seems to run in the north-south direction and covers an area of 24563 square metre with an altitude of 2.2 metre from the surface level of adjoining fields. It was occupied for the first time in late Harappan phase and thereafter remained in possession during PGW and RW levels.

The soil is composed of loam and clayey soils with a slopping tendency towards south-east. The nature of soil water in shallow aquifers is reasonable mainly because being close to the canal, but in the deeper levels it is of general nature, i.e., saline. There is a shallow depression to the north-east of the mound with clayey soils and the thin vegetation cover is also of general nature.

15. Bhawar – II: H and LH 29° 11' N 76° 30' 05" E

This lies 2 kilometre west of Bhawar village across the Sunder Canal Branch and east of the road leading to Nizampur from Bhawar. The present condition of the mound is dilapidated caused partly by the irrigation department and partly by road construction agencies. There are deep cuttings all over the site. And the elevated portions have been removed by the said earth diggers reducing it to the surface level and at some places even lower. The earliest most cultural remains represent Harappan levels. This was followed by late Harappan remains in the form of pottery and pieces of faience bangles. A significant feature of this mound is traces of a mud wall lines at certain untouched places on the surface itself. It covers roughly an area of 12000 square metre with a bare remnant height of only 0.6 metre at certain places. The longer axis are north-west—south-east oriented. And the natural conditions are more or less akin to the foregoing settlement.

16. Bichpari – I: LH 29° 12' N 76° 40' E

The ancient mound is locally known as 'Sammat-Wala' and lies one kilometre west of Bichpari village in the fields to the northern side of *kuccha* passage leading to site. The site is completely removed and transformed into agricultural fields of aeolian sandy soil. Only a few late Harappan potsherds and pieces of faience bangles were recovered from this site establishing its cultural affinity. And the spread of pottery brings out its north-west to south-east orientation. The area covered by these remains is measuring about 990 square metre contrary to its previously reported big size.[4]

Regarding surrounding micro-level natural conditions mention may be made of soil composition consisting of sandy of rising nature bearing typical vegetation of *sarkanda* and *dab*, level loam with alkaline trees, saline patches with small thorny bushes and low-lying clayey soils. The nature of subsoil water is once again not good in deeper aquifers and a limited amount of workable water is available in shallow stratum in vicinity of canal.

17. Bichpari – II: PGW 29° 10' N 76° 45' E

The ancient site is situated to the south-east of Bichpari village, east of Gohana-Jagasi Road on an elevated aeolian sandy place, which is now under intensive cultivation. The Bichpari Minar runs east of the site in north-south direction. The spread of cultural remains over an area of 2485 square metre also points to the longer axis in the same direction. Its cultural affinity is disclosed by the presence of PGW sherds. In relation to any cultural depositions we may hint in the negative.

The natural conditions around the site present a rich combination of level loam soils in the west with a sloping tendency in the same direction, which is, at local level, intervened by north-south extending aeolian sandy low ridge receding in all directions gradually and ultimately succeeded by a vast depression in the south containing clayey soil. This forms a local reservoir to receive rain water from adjoining areas and subsequently retains it for quite some time. The nature of the soil water is not good below 30-40 feet in the area. The natural vegetation is of dry character.

18. Bilbilan: RW and E. Med 29° 00' N 76° 47' E

This is located to the north-east of village Bilbilan on an aeolian sandy mound at 0.8 kilometre. Locally this is known as 'Thali Tiba'. The ancient remains are in a disturbed condition for being under cultivation. But the scatter of RW shapes clearly marks its cultural association with early historical and early medieval times. The area of circulation of these remains is 27336 square metre without any cultural deposition. The micro-environment does not appear very healthy in the absence of any perennial source of fresh water. But the very composition of the surrounding soil and the existence of a shallow depression to the south calls for rational analysis.

19. Busana Kalan: LH-OCP and GW 29° 14' N 76° 44' 04" E

This site is located east of the village and embraces the western and southern parts of the village pond, which exists there. As the site is close to the pond the area of the mound is rammed by movements of both human beings and animals. Therefore, the surface of the almost level site appears to be compact and hard. Some pieces of faience bangles, long-barrel-shaped faience bead and potsherds of late Harappan and OCP are among the cultural items to attest to its antiquity. Most of these items are found embedded in the surface soil over an area of 990 square metre. The OCP from this site bears close affinity with that found from Barota Thali. Pottery remains are still more interesting, since certain instances bear grey and yellowish colour in lower and upper parts respectively and vice-versa. This is very significant phenomenon and needs special attention.

Geographical composition is also significant because in the west, soil is sandy, loam and hard clay with uneven surfaces, whereas in the east it appears to have been a seasonal flood plain, locally known as 'Jheel' or 'Dahar'. Here the soil is sandy, grey replet with growing salinity and lacking in vegetation. It remains, in fact, fallow for its being flood prone. And flood waters used to come from Bhadauti and Jawahara villages in the north-west. Water is deep-seated and not fit for being used. But some respite have been provided by canal system.

20. Butana – I: LH and GW 29° 10' 02" N 76° 37' 06" E

The ancient site is located 3 kilometre south of the village Butana beside a *kuccha* road and locally the settlement is known as 'Banua-Wala-Khera'. Now it is in bad shape and occupies a saline soil pocket. The spread of pottery suggests its north-south orientation covering approximately 990 square metre. In relation to cultural deposition there is hardly any elevation for the mound as it is almost level to the ground.

The land adjoining to the site is salinity prone and devoid of fresh subsoil water. There are, of course, some low-lying depressions around the settlement, which might have played significant role in the life of people in the past. The overall vegetational composition mainly consists of drought and salinity resistant species.

21. Butana – II: H and LH 29° 12' N 76° 37' 02" E

This is the second ancient settlement in the revenue parametres of village Butana. And locally the site is known as 'Bania-Wali' situated one kilometre north of Butana village. The presence of Harappan dish-on-stands and perforated potsherds explains its first occupation by the Harappan people and succeeded by the late Harappans as pointed out by pottery, faience objects, etc. The present condition of the mound is not good as it has been converted into agricultural field. Its size is 1090 square metre, which is based on expansion of pottery. While it used to encompass an area of 24281.13 square metre as previously reported by Silak Ram.[5] Its extension seems to have been in the north-west to south-east direction.

Its catchment area does not look very pleasant, especially in the summer. But in rainy and winter seasons it wears a very appealing look. And in years of normal rainfall it becomes still more attractive and could prove lucrative for the rich combination of rising aeolian sandy, level loam and clayey low-lying soils around.

22. Chhapada: H, LH, PGW–BSW and RW-NBPW 29° 13' N 76° 31' 03" E

This stands one kilometre north-west of Chhapada village beside a local *kuccha* road. It is known by various local names like, 'Tiba', 'Reda' and 'Kakereda'. The entire mound is badly disturbed by saltpetre making activities. Even so it is not difficult to ascertain its cultural association and hierarchy, which travels back to Harappan times from RW-NBPW phase. It seems that it remained in existence during these periods. The representative items of the respective periods were gathered from various points of the mound, which are discussed separately in a subsequent chapter on material remains. Now it covers an area measuring 4046 square metre against earlier reported size of 4800 square metre and rising to a height of 1.6 metre as against 1.82[6]. If we discuss the geographical conditions around the site, we are bound to notice, obviously advancing salinity debilitating the soil fertility, gradually slopping tendency of surface soil from north to south (accordingly carved out a drain through a depression in the east) and the saline nature of soil water. All these features present a dismal picture now.

23. Dhanana: PGW 29° 06' N 76° 30' 04" E

The ancient settlement of Dhanana is situated west of the Lakhan Majra-Rindhana Road. Its present condition is not in good shape for its being under cultivation, yet, the sprinkle of PGW clearly points out its cultural association. It was, of course, difficult to get its precise size. The overall surroundings are not good in any manner[7].

24. Gangana – I: LH 29° 14' 08" N 76° 35' 09" E

Locally this is known as 'Huda-Wala-Kheda', which stands 1.5 kilometre north-east of village Gangana and only 300 metre west of Bichpari Minor. It belongs to late Harappan phase as indicated by dish-on-stand with dull-drooping blunt rim and a variety of faience bangles. Although the mound is now under cultivation, yet, its size is managed with the help of scatter of pottery, which comes to 4020 square metre on the level aeolian sandy soil.

The area around this settlement is provided with basically aeolian sandy soil, but pockets of low-lying clayey soils are also visibly prominent in the north-west zone where a natural depression still exists at a walking distance. In the western section appearance of saline soils, though limited, is uncomfortable. The hydrological regime is also not conducive in any manner, since quality of water is usually saline. Even the recent introduction of canal system fails to bring any quick tangible transformation in that respect. The vegetational cover is also in complete conformity with the extant conditions.

25. Gangana – II: H and LH 29° 15' N 76° 37' E

This lies 2 kilometre north of the village Gangana alongside a *kuccha* road. Locally it is called 'Kali-Khera'. The present condition is not in healthy state for being under cultivation. The entire mound is levelled by farmers. Hence, there is no question of any cultural deposition. But, even then its longer axis seem to have run in the north-south direction as indicated by extension of pottery of Harappan and late Harappan cultural strands. The rough estimate about its size suggests 4020 square metre covered area. And regarding topographical and ecological conditions it can simply be placed on record that it is akin to that of the foregoing settlement as both the sites are in close vicinity.

26. Gangesar – I: EH, H and LH 29° 11' 09" N 76° 40' 08" E

The ancient settlement is known as 'Kheri' by the villagers and stands at a distance of 0.5 kilometre to the east of the village Gangesar in the fields composed of gradually rising aeolian sandy soil. Being under cultivation it is difficult to assess its precise covered area and thickness of cultural debris. Hence, the covered area of 4020 square metre is based on estimate. Its cultural affinity is indicated by nail-headed Harappan dish-on-stand and blunt rimmed dishes and faience objects. Similarly, short-projected rims and thin section of ceramic wares point to the early Harappan character.

The surrounding fields are sloping imperceptibly to the south-east from the north-west. To the east of the site there is a shallow depression with similar orientation, which is now occupied by the Drain No. 8 leading to Gohana. Subsoil water-table is deep-seated ranging from 20-25 feet, which is now experiencing a rising tendency. But the quality of water is of regional nature, i.e., saline, though in some pockets, primarily along the canal system, working quality of water is available in only upper limits for wells and tube-wells. The entire area would render hopelessly dry, but for occassional availability of meagre canal and tube-well water. The natural vegetation is of dry nature and sparse character suiting to the prevailing natural conditions.

27. Gangesar – II: LH, PGW–BSW 29° 11' 06" N 76° 40' 09" E

This site is 1.6 kilometre south-east of the village Gangesar and about a kilometre from the previous settlement. It is called 'Badi-Kheri' by villagers. This is also partly disturbed these days and seems to have been occupied sometime in late Harappan times as suggested by occurrence of pottery of this period and frequent pieces of faience bangles and stray beads also alongwith a few sherds of BSW. It encompasses an area of 4020 square metre with north-west to south-east orientation in conformity with the flowing water course in the east. It rises to one metre and the site is situated on a natural mound, but surely of clayey soil unlike that of the foregoing site. In regard to the natural conditions these were more or less similar to those of the previous site.

28. Ghadwal – I: EH, H and LH 29° 10' 04" N 76° 32' 02" E

It stands 1.5 kilometre south of the village Ghadwal and to the west of the *kuccha* road leading to Rindhana. Generally villagers call it 'Khera' and occasionally 'Bada Kheda' also. Nowadays it is under cultivation and badly disturbed. It seems to have been established some time in early Harappan times and remained in existence till the late Harappan levels as suggested by cultural artifacts. It covers an area of 20060 square metre and rises to one metre from the surface level. It is said to have encompassing an area of 40468.56 square metre and rising to 3.04 metre from the surrounding surface level[8]. It seems that these remains are existing upon an elevated loam soil, which is now affected by growing salinity.

To the south and south-west of the mound there exists a sharp and wide depression. Before appearance of the canal-Kalanaur Branch across this depression flood water used to come from Ladana in Jind district to Nizampur and Bhawad in district Sonipat. This line of flood water course, extending in the south, is very interesting in relation to these settlements and this will be dealt with subsequently. The soil is composed of level, elevated and low-lying stretches with saline subsoil water and thin vegetation of dry character.

29. Ghadwal – II: H and LH 29° 09' 07" N 76° 32' E

This is another settlement in close vicinity of the previous one and stands about two kilometre in the south from the village Ghadwal. Generally it is called 'Kheda' by the villagers. Its southern parts have been disturbed completely by farmers, but even then the remaining parts are considerably large encompassing an area of 19000 square metre. The cultural deposition rises to 1.4 metre at present with growing slope towards the natural depression in the north, running in north-west to south-east direction, which seems to have inspired extension of habitation on the same pattern. It seems that the ancient remains are located on a natural aeolian sandy rising place. In connection with its cultural association potsherds of Harappan and late Harappan phases are scattered along with other cultural items on the surface. Keeping in view its location, being in close vicinity to other early Harappan sites, cultural thickness and size of the mound there is strong possibility of its being established some time in early Harappan times as well, but despite best efforts this could not be ascertained on the spot. So far as the geographical conditions are concerned presence of low-lying soils in three directions, except south, is quite interesting. Other factors are the same as mentioned above.

30. Ghadwal–III: EH, H, LH, PGW–BSW 29° 12' N 76° 35' E

This ancient site is situated 1.5 kilometre south-east of the village Ghadwal on almost a level soil. Locally it is known as 'Pana-ala-keda' or 'Jogian-ala'.[9] The covering area is also reduced by encroaching agricultural fields from almost all the sides and now it is reduced to only 990 square metre area. Its antiquity goes to early Harappan times and continued till the PGW levels as demonstrated by pottery shapes and other cultural artifacts. The longer axis are east-west oriented perhaps inspired by prevailing natural conditions, as the shallow depression runs almost in the same way to the south of the site.

The adjoining soil types present a very rich combination of aeolian sandy rising soils in the north, gently sloping soils of loam character in the west, depression with clayey soils in the south and again low pockets in the east. The nature of the soil water is undoubtedly not good in any respect and the vegetation of dry character is very sparse. But location of a multi-culture settlement is quite significant and calls for an analytical approach for an objective assessment of life-styles in the remote past under these conditions.

31. Ghadwal–IV: RW, Kushan and E. Med 29° 12' 02" N 76° 35' E

This ancient site is adjacent to the previous site in the north and exists on an elevated aeolian sand mound[10], which is now under cultivation and there is hardly any stratified deposition intact. The cultural remains, primarily consisting of pottery, are littered over a large portion of the sand-dune. If we go by the spread of pottery with certain deductions it covers an area of

40962 square metre. The geographical condition is more or less the same as mentioned above regarding the previous settlement.

32. Ghilaud Kalan: LH, GW and RW 29° 00' 20" N 76° 39' E

This settlement is situated on an aeolian sandy ridge of local nature, running in north-south direction and lies to the west of Ghilaud Kalan village at a distance of one kilometre. The site is completely disturbed by agricultural activities, yet, the ancient remains are quite obvious as the area is littered with pieces of RW. Late Harappan potsherds and faience pieces along with GW sherds were gathered from the south-western periphery.

To the east of this site the soil is increasingly sloping in the east to assume shape of a natural depression, which runs parallel to the ridge. This is now occupied by a local drain to take off excess rain water from the catchment area. In the west also the surface level of the soil is increasingly receding and the nature of the soil also changes to fine and fertile loam. But in the absence of suitable subsoil water and any perennial river the overall geographical picture does not portray any congenial situation.

33. Ghilaud Khurd: LH and PGW 29° 03' N 76° 39' E

These ancient cultural remains are existing on an elevated sandy soil only 600 metre east of Ghilaud Khurd village on both the sides of an approach road branching off from the Gohana-Rohtak Road and leading to the village. It is known as 'Lala-wali-Thali' and is under cultivation. But its approximate size comes to 4335 square metre without any cultural deposition. The cultural history is demonstrated by scatter of late Harappan pottery, pieces of faience bangles and PGW sherds.

The soil is aeolian in character and spreads around the settlement and witnessed sloping down in the south-west ultimately converting into a depression. Similarly to the north-east there exists a similar depression, but the subsoil water is brackish. In the summer season it appears very dull, dry and hopeless. But as soon as it experiences pre-monsoon showers, things change very quickly for the pleasant. And by the time monsoon assume proportions it becomes quite pleasant all around and it is said by the villagers that in years of heavy rainfall flood water used to come from Rukhi village via Ghilaud Kalan in the north to join ultimately the depression, which now forms the main Drain No. 8. All these factors constitute a very interesting picture.

34. Gohana–I: LH-OCP 29° 07" N 76° 04' 02" E

This ancient site is located one kilometre south-west of Gohana on the north-western fringe of the sandy mound. Cultural remains in the form of late Harappan-OCP, potsherds and pieces of faience bangles were picked up from the diggings of canal, an extension of Khandrai-Bichpadi Minor. Thus, the discovery of this buried site is a matter of chance. Hence, the precise measurement of the cultural expansion was beyond comprehension. However, the frequency of cultural objects in the exposed sections and their extension point to its small size. If one looks at the prevailing geographical condition, especially in the summer season, the emergent picture does not seem to differ from that of the general character of the zone. But the existence of the famous natural depression at about one kilometre in the north provides promising condition so far as availability of water and game are concerned. However, the subsoil water is deep-seated and saline.

35. Gohana–II: PGW 29° 12' 08" N 76° 42' 02" E

This ancient site representing the Painted Grey Ware phase is situated to the south-west of Gohana Town on the aeolian rising sandy soil.[11] These days it is under intensive cultivation. However, sherds of the PGW are witnessed strewn over 990 square metre area to attest to its cultural background.

36. Isapur Kheri: LH and GW 29° 12' 06" N 76° 39' E

These ancient remains were found on a sandy mound at a distance of one kilometre in the north-east direction from village Isapur kheri along a *kuccha* road. It is called 'Kala-Bad-ali-Thali'. Late Harappan and GW sherds were collected from an area of 4020 square metre. The physical condition of the mound is not in a healthy position as it is now under intensive cultivation, but the spread of pottery highlights the north-west to south-east orientation of the site.

The surrounding area is almost plain with a gentle slope to the south-east. The soil is loam in nature with saline subsoil water. A natural depression called 'Dahar' exists south-west of the site with north-west to south-east extension.

37. Jagasi: RW and E. Med 29° 15' 01" N 76° 40' E

This ancient mound is situated one kilometre east of the village. It is partly under cultivation and partly disturbed by salt-petre making activities. As a result everything is in topsy-turvey state. Nevertheless its cultural association is demonstrated by the presence of early historical and early medieval red ware shapes. Its approximate size is 40962 square metre. The natural environment is not conducive at all.

38. Jasrana: RW and E. Med 28° 58' N 76° 02' E

This ancient settlement is situated 2.5 kilometre north-west of the village on an aeolian sandy soil, which is now under intensive cultivation. Hence, there is hardly any sign of cultural deposition but, even then the occurrence of RW pottery shapes and terracotta objects not only highlight its cultural association, but also delineate its area of occupation, which comes to 32336 square metre, though this is based on the spread of pottery. The line of extension seems to have been north-west to south-east oriented. Away from the site the soil is either level clay loam or low-lying clayey which, very often, is vulnerable to encroaching salinity. This is mainly because of excessive salt content in the soil, saline soil water and rising water-table for waterlogging conditions and improper care by farmers. To the south of the site Drain No. 8 runs through the natural depression. The vegetation of this locale is of regional character.

39. Jawahara–I: LH and PGW 29° 00' 45" N 76° 47' 20" E

This ancient site is known as 'Gark-Kheri' and stands about 400 metre west of Jawahara village and of the drain. Nowadays the cultural depositions have been removed in order to convert it into agricultural fields. Anyway sherds of dull dish-on-stand, faience bangle pieces and PGW were visible to suggest its antiquity. These remains cover an area of 4020 square metre and there is hardly anything in the name of cultural deposition.

The nature of soil around the site is clayey of light grey colour signifying constant visit by rain water. This point is further borne out by declivity of surface level from the north-west to the south-east and the existence of a drain with similar orientation leading to Kasanda village in the south-east direction. The water-level is not deep, but it is of brackish nature. This illustrates the higher salt content in the soil, which is further manifested in the form of salinity affliction of soil. The character of vegetal cover is in full agreement with the prevailing geographical and ecological conditions. Therefore, the area wears a very dull appearance in summer and years of low rainfall.

40. Jawahara–II: RW 29° 01' N 76° 48' E

This is another site in the revenue limits of this village, which is known as 'Purana Gaon', situated east of the village and completely disturbed by farmers. A sprinkle of RW of early historical period marks its cultural association. Its area seems to have been 18980 square metre as per the spread of pottery.

41. Kahalapa–I: LH 29° 07' 03" N 76° 35' E

The remains of this settlement were marked on the aeolian sandy mound at a distance of one kilometre, north of Kahalapa village. The area is called 'Nai-ala-Khera' locally. These days it is under cultivation and late Harappan pottery, pieces of faience bangles and terracotta cakes were picked up from here. The area of circulation of these objects is about 4918 square metre. Its major ends are east-west oriented.

To the south of these remains the land is witnessed sloping in a devouring manner turning into a shallow depression, which usually receive rain water from the adjoining rising sand dunes. In the further south is level loam of very fertile nature if provided with fresh water. The nature of the soil water is not good and the water-table is deep-seated ranging in between 22-27 feet. During the summer the entire area adopts a very dreadful and unsupportive look, but, when again the area was visited in the rainy season the millet crops made it very pleasant. This may help explain the significance of micro-environment in the life of people.

42. Kahalapa–II: H and LH-OCP 29° 08' N 76° 36' E

This is another site in the revenue record of the village Kahalapa and stands on the elevated aeolian sandy pocket 1.5 kilometre north of the village, which is locally called 'Badi-Thali'. These days it is also under cultivation and the ancient remains are in a very bad shape. Harappan and late Harappan potsherds are scattered over the area. Some circular terracotta discs with finger impressions in the centre have also been picked up. Similarly pieces of faience bangles were collected. These artifacts were spread over an area of 4020 square metre but without any tangible signs of cultural deposition. Its geographical condition is in consonance with that of the previous settlement, since both these sites are in close vicinity and bear similar surroundings. Millet (*bajra*) and gram (*chana*) crops are very popular in this area, being most suitable to the prevailing dry and unfavourable climatic conditions.

43. Kailana Khas: GW, RW and E. Med 29° 10' 01" N 76° 46' E

The ancient site represents GW, RW and early medieval wares. It is situated on the northern periphery of the village Kailana Khas on an elevated loam clay mound. Locally known as 'Reda' and nowadays it is in a dilapidated position as a result of 'saltpetre' extraction. But the spread of RW sherds betrays its cultural identity. It covers an area of 32018 square metre and rises to 1.6 metre and at some places even higher. Silak Ram reported its size of 40468.56 square metre with 4.57 metre height.[12] And he associated the site with medieval period alone. A part of the mound is still encroached upon by the villagers.

44. Kakana: LH, PGW and RW 29° 08' 20" N 76° 49' E

It is one kilometre north-east of Kakana village and near a drain on the south. The mound is locally known as 'Thali' or 'Purana Gaon'. The physical condition of the mound is not good, since it is completely under freguent cultivation and with the result the cultural depositional process is not visible. The ancient remains are found scattered over a vast area, which is measured to 8134 square metre. The height of the aeolian sandy mound comes to 1.4 metre from the surrounding ground level. The orientation or extension of cultural remains seems to have been in the north-west to south-east direction conforming once again to the general sloping tendency of the surface level and consequently the flow of rain water. This is further supported by the existence of a sharp depression to the north-west, north and east directions heading towards north-east to embrace Sargthal and Nyat villages and coming from Khanpur and Samdi villages in the north-west.

The soils in the surroundings consist of level loam stretches in the south, low-lying clayey in the depressions and aeolian rising sandy soils in between. Chemically these soils bear a very high percentage of salt causing brackish to saline nature of average soil water and the water-level is, of course, well within manageable limits, i.e., 4.5 to 6.2 metre deep. The overall nature of vegetation is in line with the prevailing geographical and ecological conditions. As a result in the dry summer season the area wears a very disappointing or mecnaing posture threatening the very existence of life with its uninterrupted continuation, but timely commencement of rainy season converts it into a lively place to provide a lease to life.

45. Kathura–I: H and LH 29° 05' 02" N 76° 35' 02" E

The ancient settlement is situated 0.5 kilometre east of the village Kathura on rising aeolian sandy soil and it is called 'Kathuri-ala-Khera' locally. The cultural association goes back to Harappan and late Harappan as potsherds of these cultures are found strewn over an area of 4020 square metre. At present these remains are not in their natural position as they are disturbed by constant agricultural activities. And interestingly these remains are found located near a natural depression in the south-east direction with north-west to south-east orientation. It used to receive rain water from villages such as Dhanana, Chhapada, Kahalapa, Ahulana and Bhanderi lying all along it in the west and north-west. The orientation of the ancient remains is in the north-south direction according to the longer axis of the depression here.

The details of surrounding environment are clearly delineated by existing prominent soil types, their declining tendency, vegetation cover and nature of hydrological regime. The realm of soil types is primarily dominated by rising aeolian sand interrupted by internecine levels and shallow depressions with relatively deep water-table of 3 metre. The nature of the soil water is brackish and during summer season it further becomes a rare commodity. But in the subsequent periods (rainy and winter seasons) the situation is fairly transformed to turn it into an attractive locale. And moreover the existence of a depression in close vicinity further adds to its liveliness, which produces good crops, even today in normal rainfall conditions.

46. Kathura–II: RW 29° 00' N 76° 30' E

This is situated on the Gohana-Meham Road at a distance of about 1.6 kilometre east of the village. It is now under intensive cultivation and the whole range of ancient deposition has been completely levelled, yet, sprinkle of RW of early historical period betrays its cultural history unmistakeably. Earlier its size was 40468.56 square metre and its height was 4.57 metre as reported by Silak Ram.[13]

47. Khandrai: RW and E. Med 29° 09' 05" N 76° 41' E

It is situated on an elevated aeolian sandy mound to the south-east of village Khandrai at 0.5 kilometre distance. The Bichpari Minor runs across the eastern periphery of the mound and these days it is in a very bad shape as being under constant cultivation. As a result there is no sign of systematic cultural deposition. But the greater frequency of RW and terracotta objects along with remains of early medieval pottery highlights its cultural background. And similarly expansion of these remains not only brings out its north-south orientation (in consonance with local slope of soil), but also the area of its influence covering 38218 square metre. The average elevation of the rising natural mound is 1.6 metre from the adjoining areas.

The nature of the soils around is very significant because to the east of the village we witness extension of the Gangesar sand in the north-south direction, which dips in the east and west directions across the canal. To the west and immediate north of the village there exists a vast natural depression enlarging in the further west and containing not only fertile clayey soil, but also reasonable water in the upper strata. This yields bumper crops these days. The nature of vegetation is basically dry and thin. In the east alongside the canal nature of soil water is also of working order.

48. Khanpur Kalan–I: LH, PGW, RW-NBPW 29° 10' 20" N 76° 47' 30" E

The ancient mound is situated one kilometre north-east of Khanpur village and beside a local drain in the south. This is locally known as 'Kheda' and contains late Harappan remains in the form of pottery and faience objects restricted to south-western edge of the mound. PGW sherds are scattered all over the mound in varying frequency. And RW pottery of historical phase is also available in the northern half. Thus, the settlement represent multi-cultural strands. In regard to its physical condition it can be stated that the mound is partly disturbed by farmers from all sides and partly by digging activities of villagers in the central portions. Despite these destabilizing attempts the mound still retains 1.6 metre height from the surrounding surface level and encompasses an area of 38134 square metre with north-west to south-east axis. Its earlier reported size is 40468.56 square metre and its height was 4.57 metre.[14]

The nature of soil water is not supportive in any manner since it is brackish to saline and 3 metre deep. The soil composition is rich as aeolian sandy, fertile level loam of grey shade and clayey soils in the low-lying southern area are noticed with gently reducing nature from north-west to south-east, touching Samdi, Khanpur, Kakana, Kasanda, Kasandi, Nyat and Kheri-Damkan villages.

49. Khanpur Kalan–II: LH 29° 09' 06" N 76° 47' 02" E

This mound is situated 1.2 kilometre west of village Khanpur on elevated aeolian sandy pocket, which is nowadays under cultivation, but late Harappan potsherds, circular terracotta cakes and pieces of faience bangles were quite obvious around. These were gathered from an area of 990 square metre. There is hardly any sign of cultural deposition at the site. The longer axis of the site are stretching in the north-south direction along the shallow depression of local nature. As regards the natural environment there is hardly any sign of a healthy situation in absence of fresh subsoil water, consistent favourable rainfall pattern and good green cover, although the soil is fertile, if provided with sufficient fresh water. This is borne out by the fact of lush green sheet over the landscape in years of normal rainfall.

50. Mahmoodpur: RW and E. Med 29° 11' 04" N 76° 42' 05" E

This is situated 3 kilometre north-west of village Mahmoodpur and one kilometre east of Gangesar village across Drain No. 8 among level fields. It is under cultivation and there is no sign of any cultural deposition. But the remains of RW sherds of early historical and early medieval periods are spread over it, which encompasses an area of 38127 square metre.

This area is extremely dry in the summer season because water-table is deep-seated running in between 3 to 3.2 metre and the quality of soil water is also saline. Accordingly the soil is also alkaline in nature with matching vegetational thin cover. Salinity is rampant. As a result only drought and salinity resistant crops are grown here mainly depending on rain.

51. Mundalana–I: RW 29° 12' N 76° 45' E

It stands 4.5 kilometre west of the village and adjacent to Jawahara Minor in the east. People of the village call it 'Kheda'. And nowadays it is under cultivation, yet, enjoy 1.2 metre height from the surrounding surface level of fields. This belongs to RW group of early historical period as suggested by pottery types. The extension pattern of these remains is in the north-west to south-east direction conforming to the general sloping tendency of the land. There is a shallow depression in the west of the mound. The site encompasses an area of 48658 square metre.

The average surrounding land is arable and basically of aeolian nature with pockets of clayey, loam and saline soils around. The salt content is on the higher side, which affects the quality of soil water also. Hence, subsoil water in this section is brackish. And as we know, such a combination of geographical features added with dry ecological conditions support only a very sparse vegetation of alkaline nature. Therefore, the composition of vegetation is in line with the existing conditions. And the emerging scenario is very hostile in the summer season.

52. Nizampur: GW and RW 29° 14' N 76° 30' 04" E

It exists one kilometre to the north-west of the village Nizampur on an elevated aeolian sandy pocket, which is now under cultivation and is completely devoid of any cultural deposition. To highlight its cultural association potsherds of GW group and RW pottery types of early historical period are frequently available from an area of 22138 square metre. The spread of pottery exhibits its north-south orientation.

The surroundings of this settlement are not very attractive because of gradually rising aeolian sands, deep-seated saline water-table and non-existence of any permanent source of natural fresh water. These factors promote only drought and salinity resistant plants in a sparse manner. But surely with the onset of monsoon season the hostility is drastically reduced to make it conversely pleasant.

53. Nuran Khera–I: H and LH 29° 11' 30" N 76° 35' E

This is situated 2 kilometre south-east of the village Nuran Khera and locally known as 'Bandari-ala-Kheda'. Being under cultivation its physical condition is deteriorated. But its cultural identity can easily be established with the help of Harappan and late Harappan pottery types and terracotta cakes. These objects are spread over an area of 4020 square metre. In regard to extant environment there is nothing special except existence of a small natural depression to its west.

54. Nuran Khera–II: LH, PGW-BSW 29° 13' N 76° 33' E

This settlement is 2 kilometre east of Nuran Khera village, one kilometre south of Gohana-Jind Road and 1.5 kilometre on the left side of a minor canal north-east of Janata College Butana.[15] These remains are found on the rising aeolian sandy mound, which is locally known as 'Gadele-ala-Khera'. Nowadays these remains constitute an agricultural field. Its cultural antiquity is marked by availability of sherds of late Harappan dish-on-stand, faience objects and large circulation of PGW, which cover an area of 4020 square metre. Earlier its size was reported about 32374 square metres.[16] Its expansion is noticed in the north-south direction and a natural depression with similar orientation lies in the west at a distance of 400 metre. The depression is now obliterated by farmers, but primarily by canal running across it. It seems that before the appearance of this canal the depression would have assumed considerable proportions in the monsoon season by accommodating rain water from catchment areas which, thus, would have been retained for most parts of the year, if not year round.

55. Nuran Khera–III: EH and LH 29° 11' 30" N 76° 34' E

The ancient remains are lying 1.7 kilometre north-west of village Nuran Kheri alongside a local passage. These days the site is fully under cultivation. Locally it is called 'Tuhia-ali-let'. The early Harappan and late Harappan remains embrace a small area of 4020 square metre with north-west to south-east axis as pointed out by spread of the objects. The natural environment is akin to that of the previous settlements.

56. Rindhana–I: EH, H and LH 29° 08' N 76° 31' 10" E

This is situated 2 kilometre north-east of Rindhana village

across the Kalanaur Canal Branch. It is locally known as 'Bada-Kheda' or 'Singhadua-ala'. It has been truncated from almost all the sides, but especially from the southern direction a major portion is removed by farmers. And now it contains 4020 square metre area with an altitude of only one metre. It has been reported that priviously it had been enjoying a size of 16197.42 square metre and rises to 1.82 metre in height.[17] The longer axis run from north-west to south-east as per orientation of the natural depression in the north. This depression assumes dimension in the east, where, even at the time of exploration a good quantity of water was available despite the ongoing dry spell of 1987.

57. Rindhana–II: EH, H and LH 29° 08' N 76° 31' E

This is just in the west of the previous mound and is known as 'Chhota Khera'—small mound. It is also disturbed from all sides and the remaining portion contains potteries of the aforesaid phases along with elliptical and triangular terracotta cakes. The height of the mound is 1.6 metre from the surrounding fields. And the depression lies in the north.

58. Rindhana–III: EH and LH 29° 08' 30" N 76° 31' 45" E

This is another site in the revenue limits of village Rindhana and is located 2.5 kilometre north-east of the village, east of canal and 0.5 kilometre north of the big mound. Locally people call it 'Kheri', which is now under cultivation and encompasses an area of 12335 square metre with 0.6 metre elevation. The cultural association is highlighted by occurrence of early Harappan and late Harappan pottery types, terracotta cakes and pieces of faience bangles. The surrounding soil is plagued with encroaching salinity, which reduces it to a very poor landscape.

59. Rindhana–IV: LH 29° 07' 50" N 76° 32' 20" E

This is another settlement in the revenue limit of village Rindhana and lies at a distance of 2 kilometre east of the village on the left-hand side of Gohana-Rindhana Road, just 300 metre inside. The present condition of the mound is not good for being under intensive cultivation. The late Harappan remains cover 8146 square metre area. To the south of the mound there is a shallow depression extending increasingly towards south. Soil suffers from salinity, but millet crops are found all around in rainy season.

60. Rindhana–V: PGW 29° 07' 08" N 76° 31' 48" E

This mound exists only 200 metre south of Rindhana-I and is locally known as 'Kikar-ali Kheri'. It is a single culture settlement representing only PGW. And the remains are in a disturbed state, since the entire mound is levelled by farmers. PGW sherds are found spread over an area of 4020 square metre. The surrounding soil is of grey shade and hard nature approaching to clayey soil, though it is level area. Millet crops are very popular for their suitability.

61. Rindhana–VI: RW and E. Med 29° 11' 05" N 76° 30' E

This settlement is situated south-west of the village at half a kilometre distance. It is also under intensive cultivation and RW pottery shapes are frequently available from a large area of 36142 square metre with an elevation of 1.3 metre from the surrounding surface level. Here the soil is aeolian sandy with saline water and affliction of salinity. To the east of the mound exists a shallow depression receding gradually in the east.

62. Rithal: LH, PGW–BSW 28° 55' N 76° 40' E

The site is 800 metre north-west of village Rithal on a local passage and is locally known as 'Majare-ala'.[18] It is in a dilapidated state, though PGW sherds were frequently available to indicate its cultural identity, which is further extended to late Harappan times by a copper bead and large number of faience bangles. These remains are stretched in the north-west to south-east direction. The thickness of the cultural deposition is difficult to establish for their being disturbed, but villagers recall of its big size and height before destruction by earth-diggers, though at present its size is only 4337 square metre.

Its surroundings are not pleasant as salinity is on the higher side, subsoil water is not good and the area suffers from acute waterlogging. There exists a vast 'dahar' or depression in the east and south of the village which gives an impression of an oval shaped vast lake. This receives rain water from surrounding areas to retain it year round. And to protect the present habitation from encroaching rain water people had to construct an earthen bund in 1978, encircling the village. And similarly the flood threat might have also loomed large over the people in the past. But at the same time this lake might have also served as the main lifeline.

63. Riwara: RW and E. Med 29° 01' N 76° 44' 04" E

This settlement lies one kilometre west of the village and just 200 metre north of Drain No. 8. It is completely disturbed by farmers and primarily by saltpetre extraction. Nowadays it is in a dump condition. Hence, the systematic cultural depositional process, expansion and orientation is lost forever. But availability of RW pottery shapes of early historical and medieval times help to ascertain its cultural sequence. Since the site is in the heart of the Bhangar the environment is of regional character. In relation to micro-environment we can add existence of a natural depression harbouring now Drain No. 8 in the south and presence of rising sandy soil in the north-east.

64. Rukhi–I: EH 29° 03' 45" N 76° 39' 60" E

The ancient site is situated 2 kilometre west of the village on the right hand side of Rukhi-Chhichhadana Road.[19] It was found on an elevated sandy soil, which now forms part of a field. Potsherds of early Harappan class were picked up from the surface, which cover an area of 2040 square metre without any cultural deposition. It was difficult to ascertain its longer axis because of its disturbed condition.

If we want to have a brief look on the existing micro-geographical and ecological conditions a reference of a shallow depression to the south with a slope towards south-west, aeolian soil composition with relatively deep-seated water-table of saline quality and sparse nature of drought resistant plants are noteworthy. And these conditions support cultivation of drought resistant crops needing minimum cultivation, labour and input. Hence, millet crop is most popular during rainy season.

65. Rukhi–II: LH and PGW 29° 02' N 76° 41' E

This settlement was found on the southern extension of the present village Rukhi among the fields on almost level surface. But in the past the site enjoyed height as reveal by villagers, which was subsequently levelled in order to make a field. As a result its original shape is lost forever. But scatter of pottery is still quite prominent, which highlight a composition of late Harappan thick sturdy red ware and PGW. Besides faience objects are also frequently available. These cultural objects encompass an area of 4020 square metre.

Its catchment area does not appear conducive at the first sight and that too in the scorching heat of dry summer. But on viewing the surroundings on micro-level we may encounter a shallow depression across the Gohana-Rohtak Road in the west, stretch of clayey loam soils around and big natural depression of regional nature accommodating Drain No. 8 in the east, which unfold other sides of the natural conditions. These factors might have played significant role in the life of people in the past.

66. Rukhi–III: RW, Kushan and E. Med 29° 03' N 76° 40' 01" E

This site is situated in the west of the village Rukhi at a distance of one kilometre and just west of the Rukhi railway station on Rohtak-Gohana Track. It is on the rising aeolian soil. Like other settlements this is also in a disturbed condition and the entire mound had been levelled by farmers for cultivation. However potsherds of RW, Kushan and early medieval periods are quite freely available on the surface, which testify its cultural identity. Since the remains are not in undisturbed position its covered area is based on the spread of the remains, which encompasses 24227 square metre area. In order to gauge the prevailing conditions north-westward sloping tendency of aeolian soil with relatively deep brackish soil water and thin vegetation are noteworthy. Such soils produce good crops of barley and millet with the help of nominal rainfall, minimum labour input and simple technological tools.

67. Sargthal: LH-OCP and GW 29° 07' N 76° 50' 01" E

The ancient remains are found lying on the sandy mound in close vicinity of the village in the south. This is locally known as 'Thali'. A village passage leading to Dodwa runs through the site in the north-south direction. The orientation of the remains are in the north-west to south-east position and covers 13335 square metre area. Ancient remains suggest beginning of life sometimes in late Harappan–OCP levels and followed by GW phase.

Around the mound soil is aeolian in character and fertile, if provided with water. The quality of water is not good and it is deep-seated. Wells and tube-wells are not visible in the area and even village ponds go dry in summers to cause manifold hardships to the people. This is further added by sparse alkaline vegetation.

68. Siwana Mal – I: PGW and RW 29° 00' 10". N 76° 35' 45" E

This ancient mound lies one kilometre south-east of Siwana Mal village and just west of Bhiwani Canal Branch. The entire mound is disturbed by farmers and canal department for digging earth. As a result pot-sherds of PGW and GW first noticed on the canal bank itself and subsequently collected from 4020 square metre area of the adjoining fields. It seems to have been a large settlement as demonstrated by extensive spread of pieces of pottery. The soil is clay loam and enjoy working quality of soil water, but especially in the shallow aquifers caused by canal system.

69. Siwana Mal – II: GW and RW 29° 17' N 76° 35' E

This ancient settlement stands 2 kilometre north of the village among the level, but approaching to low-lying fields. This has been disturbed by saltpetre making agencies and presently everything is in a topsy-turvey manner, which deprived it of its chronological sequence. Even then finds of GW and RW sherds underscore its cultural identity. The spread of these objects also highlights the area of their circulation, which is 27234 square metre and rise up to 1.2 metre at present. The orientation of their extension not obvious. If we take into account the general sloping tendency of the shallow soil surface and the average relationship between it and the orientation of settlements, then it must have been of the north-west to south-east order.

Coming to the existing micro-environmental conditions we may put on record the unproductive quality of soil water, which is riplet with high concentration of salt. This is also reflected in the surface soil and as a result in absence of proper care/ remedial steps the salt concentration in the form of white or greyish powder is quite visible, especially in the low-lying area in the south, which wears a waterlogged appearance.

70. Samri: LH-OCP, GW and RW 29° 11' N 76° 47' E

The ancient remains are situated 2.5 kilometre south-west of the village and only 200 metre east of Chidana-Khanpur Canal. Nowadays this is disturbed by farmers, but harbours OCP, GW and RW underline its multi-culture nature. These remains are scattered over an area of 24063 square metre of aeolian sandy nature in the north-west to south-east direction with an elevation of 0.6 metre from the surrounding surface level. Its earlier size is said to have been 16187.42[20] square metre. In the summer natural environment does not present a very healthy picture, but subsequently it turns out to be very attractive pocket though mainly on account of normal rainfall.

71. Thuska: PGW-BSW, RW-NBPW 29° 06' N 76° 40' 45" E

The ancient mound lies 400 metre south-east of village Thuska and bears PGW to suggest its inception. Its subsequent occupation is highlighted by presence of RW. These remains encompass an area of 24563 square metre, but devoid of any elevation in absence of undisturbed conditions. To the south of this site a sharp depression is quite obvious through, which a local drain to take off excessive rain-water runs. It seems that before the introduction of the drain the depression might have received rain-water from the adjoining areas to assume greater proportions. Thus, the added quantity of water would have helped to retain water for the most, if not whole, parts of the year. And such a situation might have gone a long way in the life of people at the settlement over the period of time.

Sonipat Tahsil

This tahsil covers the eastern parts of the district and is consisting of two geographical units such as the Bhangar and the Khadar occupying the western and eastern sections of the tahsil respectively. The former is relatively a small constituent than the latter. These two contrasting zones are not only adjoining in nature, but also run parallel in the north-south direction. The eastern low-lying sector is very rich for agricultural as well as pastoral activities for its being new-alluvium–a present riverbed. This belt is replete with tremendous amount of fresh water from top to the bottom and littered with depressions of varying proportions, which might have served as major lifelines in the past. There was an inherent mechanism of annual refreshment of soil fertility through seasonal inundation, which has now been blocked by a bund. These days this very belt is supporting a very large number of modern settlements of usually small size converting it most thickly populated area. There is also another trend to establish new villages, even along the active watercourse of the Yamuna river (map 4).

As compared to the Yamuna belt the Bhangar is not only higher, but also drier in nature. The soils of this zone are composed of level loams, low-lying clayey and rising aeolian sandy soils. The soils of the last category are relatively of low frequency and proportion, whereas in some pockets low soils predominate. The nature of the soil water is no better than that of the Gohana tahsil, but surely the water-table is shallow in comparison to the one witnessed in the Gohana-tahsil — the western Bhangar. Similarly the frequency of workable water in shallow aquifers is also higher than the western Bhangar. Nowadays in this section we witness a very effective network of canal system and relatively better population concentration than the western Bhangar, but low density than the Khadar in the east. Keeping in view these contrasting characteristic features of the constituent zones of this tahsil a brief description of the explored ancient settlements is attempted to in the following paragraphs.

Description of the Explored Settlements

1. Agwanpur: GW, RW and Med 29° 06' 15" N 76° 58' E

The ancient settlement is situated just south of the village and overlooks the low-lying Khadar[21] in the east. It is partly under occupation of the villagers as they store their dung cakes and dry fuel sticks. But, even, then the whole mound is more or less intact. Locally this is known as 'Khera' meaning thereby a residential place in general sense and stands for a deserted residential settlement in restricted sense of the word. And for this purpose the latter connotation is valid. The present mound covers an area of 27736 square metre and rises to 1.3 metre from the western surface level, whereas in the eastern sides it is more than two metre. A sprinkle of GW sherds of PGW group and RW pottery of NBPW period speaks of its cultural association. Sherds of glazed ware were also picked up from the surface.

To have a peep into the existing micro-environment we cannot but explain the fertile low-lying alluvial flat in the east enjoying fresh soil water very near to the surface to offer opportunities for subsistence in all the conditions. And similarly in the western section a rich combination of level and rising soils in varying degrees with reasonable soil water sounds equally healthy. The combination of these two features is quite promising to unfold various vistas for subsistence activities.

2. Akbarpur Barota: PGW-GW-BSW and RW-NBPW 28° 04' N 77° 54' E

This ancient settlement is situated to the north and west of the village Akbarpur Barota[22]. Nowadays parts of the mound are under modern habitation. Only in the north-eastern section a small part of the mound is intact. As per the gathered material it travels back to the PGW phase. And subsequently, it seems to have remained in possession during NBPW phase. However, existence of pieces of faience bangles may hint at its still anterior nature. The exact measurement of the mound is difficult for its being under occupation, yet, the spread of the habitation in different directions suggests its extension over an area of 18134 square metre. This rises to 2.6 metre at places from the surrounding surface level.

So far as the surrounding natural environment is concerned the settlement seems to have been located at a very promising point as it stands at the right bank of an incoming natural depression dissecting the Bhangar in the west and the Khadar in the east. The line of the depression is very clear for its being very deep and sharp in relation to the surface level of the western section. And equal contrast is seen in soil constituents and the nature of soil water. Thus, it could have offered ample subsistence opportunities in all conditions. The natural depression here adopts a east-west course as against its normal north-south orientation. And two major seasonal drains (Drain No. 8 and 6) join at this place and afterwards join the Yamuna river in the east.

3. Badwasani: RW, Kushan and E. Med. 29° 00' 45" N 76° 57' 48" E

These ancient remains are located on the south-eastern fringe of Badwasani village and partly dissected by Sonipat-Gohana Road. The major parts of the mound have been encroached upon by present habitation, which makes it difficult to ascertain its dimensions. But the existing remnants are indicative of its reasonably large size[23]. To the south and west of the mound exists a large depression in the north-south direction, which can accommodate a good amount of rain-water. Alongside this depression the soil is clayey in nature, which is considered very fertile to grow bumper crops. Away from this water-line the soil is composed of rising aeolian sands and level loam. Such a natural composition is full of a variety of subsistence opportunities.

4. Bahgru: EH 28° 59' N 76° 56' 04" E

The ancient site is situated 1.25 kilometre north-east of the village on a sandy pocket and about 400 metre east of another mound with medieval remains. Now it is partly under cultivation and partly under residential houses. Some potsherds with black block bands on the necks and internally incised with zigzag lines were picked up from here to establish its cultural identity and this has already been reported by Bhan.[24]

5. Bajana Kalan: RW and E. Med. 29° 09' N 76° 47' E

This lies to the north end of village Bajana Kalan and these days it is partly under modern habitation and partly under cultivation. Under these conditions it was difficult to workout its length and breadth. However, it seems to have been a big mound with east-west axis. RW potsherds are freely available from here to testify its antiquity. Around this settlement the soil is mainly aeolian sandy with occasional saline patches. Water is also not good in taste and tube-wells are successful in shallow aquifers ranging between 30-40 feet. And the vegetation is also in complete agreement with the existing inhospitable natural conditions.

6. Balandpur Kheri: PGW-GW, RW and Med 29° 10' N 76° 56' E

The ancient mound is located to the south of the village and to the north of Ganaur-Balandpur Kheri Road[25]. Regarding its cultural background it is to be noted that previously only GW was reported from the site to suggest its beginning. But, during my exploration a few sherds of PGW were also collected from the eastern sections of the mound. Thus, its antiquity is pushed back in time. Regarding physical condition of the mound it can be stated that it is under constant multi-pronged attacks from farmers, earth and brick diggers and development schemes of Haryana Government. Consequently the entire mound is obliterated and littered with brickbats of medieval times and potsherds representing various cultural stages. The site is locally known by various names as Myana, Kota, Satkumbha and Jalalabad. According to the local people it was capital of Raja Chakavabaina, whose reference in literature is not available. Its size is 202150 square metre and height is 24 metre from the surface level as reported by Dutt[26]. To judge the natural micro-environment one can easily discern existing positive features like the fertile new alluvial soil with fresh water in the east and the undulating aeolian soils in the west with suitable soil water. This type of natural combination is very promising for a wide range of subsistence activities. This seems to have played a significant role in the lives of the people through the ages.

7. Bhadana: RW and E. Med 28° 57' N 76° 56' 04" E

This ancient site is situated 1.60 kilometre south of the village and now under constant cultivation. Locally it is known as 'Reda'. Its ancient remains of early historical period are very scanty because at present the mound is completely removed. Earlier it is said to have been a big mound spread over several acres of land and rising to considerable height. But now very rich crop of wheat are grown over that.

8. Bhagan : RW and E. Med 29° 04' N 77° 03' E

This mound stands 0.6 kilometre in the south-west direction of village Bhagan, which is these days under cultivation. Its cultural identity is illustrated by occurrence of RW shapes of early historical period and similarly subsequent occupation is demonstrated by pottery shapes of early medieval period. These remains cover an area of 18236 square metre and there is hardly any elevation. But its surroundings speak of hospitable conditions. Because the soil is alluvial and the subsoil water is very good for use. Moreover in the close vicinity of the site there exists a natural depression, which might have been of greater size in the past to serve as a source of surface water to the people.

9. Bhamad: RW and E. Med 29° 08' 05" N 76° 56' 02" E

This lies to the north of village Bhamad at a distance of 0.8 kilometre and in the west of a nullah or drain. The present condition of the site is not intact because it is partly disturbed by farmers and partly by saltpetre collection activities. Its cultural association is highlighted by the presence of early historical RW shapes, which is succeeded by early medieval pottery types. Its extension seems in the north-south direction and possess 1.2 metre elevation. The overall natural environment does not appear conducive because of rampant salinity and saline subsoil water affecting growth of life adversely.

10. Bhatgaon – I: RW and E. Med 29° 59' N 76° 54' E

The ancient settlement containing RW of early historical and early medieval periods as pointed out by their scatter, is situated one kilometre west of the village Bhatgaon on the righthand side of the Sonipat-Frmana Road. It is a reasonably big mound covering 36236 square metre area and rises to 1.8 metre. These days the most parts of this mound are intact in its original position. The surrounding soil is aeolian sandy in character, which is considered rich if provided with sufficient amount of fresh water. But fresh water is confined to uppermost levels only, which seems to have been of recent origin effected by the introduction of canal system in the area.

11. Bhatgaon – II: LH, GW and RW 29° 59' N 76° 54' E

This is another settlement in the revenue limits of the same village mentioned above. This is situated about 300 metre west of the previous mound. The road has bifurcated the northern side. Similarly, a local water channel, with a well along, running in the east-west direction has also divided it into two halves. Interestingly the southern half on rising sandy mound is contalnlng late Harappan potsherds and pieces of faience bangles along with GW of rough nature. Whereas, the northern half is possessing RW of early historical period. This part is intact, but the southern section is under cultivation. The geographical background is akin to that of the previous site.

12. Bhatgaon – III: LH–OCP 29° 59' N 76° 56'E

This is situated to the east of the village at a distance of 800 metre on a sandy mound to the north of the road from Sonipat. It contains late Harappan pottery and pieces of faience bangles. At present its condition is not good.

13. Bidhal: LH-OCP 29° 03' 06" N 76° 47' 08" E

This lies about 2 kilometre south-west of village Bidhal on a rising aeolian sandy pocket, which is now under intensive cultivation. As a result there is no sign of cultural deposition except a thin sprinkle of late Harappan and OCP. Besides numerous pieces of faience bangles are also quite visible, which highlight its cultural background. But in the given situation it was difficult to ascertain precisely the area of occupation. However, the size of the site seems to have remained small like other such settlements. To the south of these remains there exists a shallow natural depression, which usually received rain-water from the surrounding areas and help to produce good 'Kharif' and 'Rabi' dry crops. This is further added by the existing rich combination of soils—aeolion sandy, level loam and clayey in different pockets.

14. Bohla: LH and PGW 29° 05' N 76° 51' 50" E

This ancient deserted settlement is situated on the north of Bohla village across the road, which is locally known as 'Reda' or 'Kallar' rising barren soil patch.[27] At present it is in a very bad shape and partly covered by villagers and partly by the Primary School building in the east and west respectively. It covers an area of 4020 square metre. Material remains from here suggest its first ever occupation in late Harappan times as suggested by pottery shapes and pieces of faience bangles. And these remains are followed by PGW. Its surrounding natural conditions are not good at all in the absence of any permanent source of water.

15. Chatia-Aulia: GW and RW 29° 04' N 76° 57' 20" E

This stands on the south-eastern extension of the village Chatia-Aulia and overlooks the low-lying Khadar in the east. Locally people call it 'Miran-Wala-Khera'[28]. The longer axis are in the north-south direction in accordance with the extension of the Khadar — the low-lying depression. Its covering area is 4020 square metre at present, but earlier its size is reported 28328 square metre by Dutt[29].

To assess effectiveness of the micro-environment we may mention existence of a sharp depression in the east running in north-south direction. This is, though obliterated at large-scale by agricultural activities, yet, quite distinct to catch anybody's sight. And this forms the dividing line of the Bhangar in the west and the Khadar in the east. Such natural combination would have offered a wide range of subsistence activities.

16. Chatia Deva: LH and PGW 29° 06' N 76° 53' E

The ancient site lies one kilometre south-west of village Chatia Deva on a rising soil of sandy loam nature. And situated east of a canal at about 300 metre away. Nowadays it is under intensive cultivation and the ancient remains have been levelled. Therefore, there is no cultural deposition at present. But the sprinkle of late Harappan, PGW sherds and pieces of faience bangles reveal its cultural identity. These remains are extending visibly in the north-south direction, that is in agreement with the existing orientation of the natural depression in the west. The area of influence is 5227 square metre.

Its surroundings do not appeal much because of prevailing saline condition around and brackish soil water. But the natural depression in the west might have served as a water reservoir to meet such requirements.

17. Chhatehara : RW and E. Med 28° 54' 06" N 77° 04' 04" E

The ancient settlement is situated one kilometre south-west of the village on an aeolian sandy mound, which is now turned into a field. Hence, there is hardly any sign of cultural deposition. Only scatter of potsherds is marked indicating its cultural association to RW of early historical phase and early medieval times. These remains are found extended over an area of 36227 square metre with north-south longer axis.

The soils of this area are composed of aeolian sand to a great extent. Subsoil water is of working nature, which is mainly confined to the uppermost stratas. In the deeper aquifers it is again of saline nature. To the east of the mound the area is low-lying, which experience usually water logging in normal rainfall conditions. This points to its significance in relation to ancient cultures. Moreover the incoming Khadar line lies at a short distance in the north-east direction across the village.

18. Chirsami: GW, RW and E. Med 29° 10' N 77° 03' E

The ancient site is located 300 metre south-west of Chirsami village on an elevated pocket. People generally call it 'Tiba' or 'Reda'–a mound. This is now partly disturbed by farmers and only a very small patch is in its natural position. The disturbed portion is full of cuttings of varying proportion. Under these conditions to recover precise dimension was not feasible, but the spread of pottery covers 18223 square metre area with an elevation to 0.6 metre from the surrounding surface level. To evaluate the existing natural conditions we may state that the average soil is alluvial in nature with sweet sub-soil water at a depth of 4.3 metre. However, occasionally saline patches are also quite visible. And in such pockets white efflorescence on the surface can be noticed. The vegetation cover is nominal, which reduces the landscape intolerably dreary in the summer season.

19. Datauli: LH, PGW-GW, RW, Kushan and E. Med.
29° 04' 15" N 77° 03' E

The ancient mound is located on the north-eastern fringe of Datauli village and north of the road from Ganaur to Bega. This is a large settlement and the longer axis enjoy north-west to south-east orientation. The area is 38723 square metre and rises to 1.6 metre from the surrounding surface level. In regard to its present condition it can be said that largely it is undisturbed except in northern and north-western sections, where minor cuttings have been inflicted. Most parts of the settlement are covered with RW of early historical and late historical periods. Only a few potsherds of PGW, GW along with pieces of faience bangles were picked up from the cuttings on the western periphery. Village women collect clayey soil from here for their domestic purposes. Thus, it was first occupied sometime in late Harappan levels.

Since the river Yamuna is not far off from here in the east and the mound falls well inside its flood plain—the Khadar, the soil is alluvial in nature and replete with fresh water. It tends to slope down towards south-east, i.e. the Yamuna river. Besides there are some ox-bow lakes, low-lying pockets and depressions around. Apart from the area of the settlement is on 223 metre contour line above sea-level as per the 'Toposheets' of Geographical Survey of India. Thus, the area being safe from the Yamuna floods appears quite promising.

20. Dhaturi: PGW GW, RW, Kushan and Gupta
29° 04' 15" N 77° 03' E

This is situated to the west and south-west of Dhaturi village[30]. At present its condition is not good for large-scale digging and encroachment by the villagers. Appearance of village huts hamper to scale the size of the mound, but it seems to have been a reasonably large settlement as its size earlier reported was 8134 square metre. The thickness of cultural deposition rises to 1.3

metre. And its longer axis are north-south oriented. The surrounding natural condition is conducive, since the site is in the Khadar plain. The general nature of the soil is plain with occasional interruptions by occurrence of elevated alluvial fans and low-lying depressions. Such a combination is pregnant with various avenues for subsistence, that is what we experience these days also.

21. Dodwa: LH and GW 29° 05' 56" N 76° 49' 45" E

The ancient remains are situated 0.5 kilometre west of village Dodwa. The mound in its major parts is undisturbed and its soil is composed of aeolian sand. It covers an area of 39236 square metre with 2.1 metre height. But the cultural remains were gathered from a very limited area and the deposition is very thin showing a sparse population. Grey ware sherds along with red pottery are visible here and there. And late Harappan potsherds and pieces of faience bangles were picked up from the cuttings in the south-western section. Thus, it was first occupied by late Harappan people sometime in that phase and followed by GW phase.

The soil of this area is consisting of level loam, elevated aeolian sand and clayey in the low-lying areas. To the north of the mound there exists a big depression, which is locally known as 'Dahar'. This receives rain-water from surrounding areas and swells in the monsoon season to assume significant proportions. This provides various opportunities for eking out subsistence in absence of fresh soil water.

22. Farmana–I: PGW–BSW 28° 55' N 76° 45' E

It stands one kilometre south-east of Farmana village.[31] Its present condition is deteriorating for its being under cultivation. Frequent occurence of PGW sherds marks its cultural association and covers 4020 square metre area. The present natural condition is not good, because the soil water is saline and soil is suffering from salinity. The only solace one can draw from the existing large depression (*dahar*) in the north, which runs almost in the north-south direction.

23. Farmana–II: EH, LH and PGW 28° 58' N 76° 50' E

This ancient settlement is situated east of Farmana village, which is now under cultivation. A sprinkle of early Harappan, late

Harappan and PGW pottery is obvious on the surface to underline its cultural identity. These remains are spread over an area of 10000 square metre only with no elevation whatsoever. The micro-environment is more or less similar to that of the previous settlement.

24. Ganaur: RW and Med 29° 08' N 77° 00' 02" E

This settlement is situated to the west of Ganaur and the road from Sonipat to Ganaur. Now it is badly disturbed and encroached upon by present inhabitants. But, even, then RW shapes of early historical period are quite obvious in the western section. There is hardly any sign of cultural deposition. Therefore, it was difficult to work out its expansion. So far as the natural environment is concerned we may mention rich combination of aeolian and alluvial soils, fresh soil water and undulating surface, which altogether offer a wide range of subsistence activities.

25. Garhi Bala: RW and E. Med 29° 50' 01" N 77° 01' E

This is situated one kilometre north of Garhi Bala village and west of Western Yamuna Canal. Locally this is known as 'Kakreda'—a place with gravel surface. It is partly under cultivation and therefore, there is not any systematic deposition of cultural material. Its cultural sequence is indicated by the spread of RW pottery of early historical period and early medieval times. The area covered is about 16234 metre as pointed out by sprinkle of pottery.

The surroundings of this site are not good looking because the average soil water in deeper aquifers is saline. Limited amount of workable water is possible in the upper levels only, which has been transformed by constant flow of canal fresh water. The soil presents different qualities like level loam, rising aeolian and saline patches. This can prove fertile if sufficient water for irrigation is provided with. In absence of this facility it wears a very dull and dreary look. This is further added by the presence of only drought resistant plants.

26. Gorad: PGW 28° 57' N 76° 49' E

This ancient settlement is situated 0.8 kilometre north-east of Gorad village and now under cultivation. Frequent PGW sherds are available on the surface covering 4020 square metre area.

There is no cultural deposition. The extension is from east to west as indicated by spread of pottery.

The soil around the site is clay loam with saline subsoil water. Leading further north the soil increasingly becomes clayey and especially in the natural depression, which now harbours Drain No. 8. The nature of the vegetation is alkaline and therefore, conforms the regional character. During summer season and in drought periods the area appears extremely inhospitable. Whereas in periods of normal rainfall it presents a very pleasant side of it.

27. Gujjar Kheri: RW 29° 09' 05" N 76° 58' 04" E

The ancient remains are located to the south of the village on an elevated mound. These days it is partly disturbed by farmers and encroachers. RW shapes of early historical period are quite obvious on the surface to mark its cultural antiquity. These remains cover an area of 38236 square metre. Since the mound is situated at the junction of the Bhangar and the Khadar people in the past might have availed of opportunities of the two contrasting zones.

28. Gumad: PGW-BSW and RW-NBPW 29° 07' 20" N 76° 58' 30" E

The ancient mound is situated east of Gumad village at a distance of 400 metre[32] and is bifurcated by the Ganaur-Khubdu Road. The present condition of the mound is not good because it is levelled by farmers in the north and east. Its size is 24258 square metre and rises to 1.6 metre from the surrounding surface in the north. Silak Ram has reported its spread over 24281.13 square metre area rising to 1.82 metre. Its cultural association travels back to protohistoric period as fine sherds of PGW can easily be picked up from the surface. Occasionally BSW is also visible. The cultural assemblage is further added by the occurrence of RW along with NBPW. The relative frequency of BSW and NBPW is of course low.

So far as the surroundings of this site are concerned we may take into account the very location of the site on the dividing line of the Bhangar in the west and the Khadar in the east, the fertile undulating dry soil of the west with brackish water and the rich alluvial soil with fresh water-table of the east and the incoming natural depression encircling the mound from the north and the east. The western section is placed 230 metre above sea-level while the eastern parts are only 224 metre. Thus, the average difference between these two zones comes to 5 to 6 metre. This situation extends tremendous avenues for subsistence.

29. Halalpur – I: PGW 28° 51' N 77° 01' E

This is situated north-east of Halalpur village on an elevated aeolian sandy soil, which is now under intensive cultivation and as a result the remains are completely disturbed. Chronological depositional process is missing. However, occurrence of PGW and GW is more than sufficient to establish its cultural identity. These remains cover an area of 4020 square metre. In its surroundings we witness extension of aeolian sand all around, which tends to slope down in all the directions imperceptibly until it is merged with level soils of sandy loam nature. The subsoil water in the shallow aquifers is of reasonable quality, but in the lower stratas it is of saline nature. The character of the present vegetation is in full conformity with the natural environment. Consequently most of the plants are of alkaline nature, which could withstand dry and saline conditions. However, in years of normal rainfall it is full of opportunities.

30. Halalpur – II: EH, LH, RW and E. Med. 28° 49' N 77° 00' E

This is another mound in the revenue limits of this village and located south-west of it on slightly rising aeolian soil[33]. This is now under intensive cultivation and the cultural remains are very thin, but sufficient to underline its cultural background. Earlier only early medieval remains were reported from here, but during my exploration I discovered potsherds of early Harappan with thick black bands of painting and wedge-shaped burnt bricks (see Figs. 5.57-59) along with early historical pottery to push back its antiquity to protohistoric period. And this collection would not have been possible from the surface, but for the ongoing digging for making bricks. An oval shaped vertical, but broken formation like a *bhatti* or *tandoor* was also noticed. The covered area is about 18236 square metre, which is primarily based on the scatter of potsherds. Its micro-environment is more or less akin to that of the previous settlement.

31. Harsana Kalan – I: LH and PGW 28° 57' N 77° 02' E

This is located 1.5 kilometre south-east of the village and north of Harsana Khurd village. The canal named Harsana Minor runs through the mound in east-west direction. The mound is disturbed by various agencies like grave makers, farmers, irrigation department, etc. And these days its western section is used as a cremation ground. Despite all these troubles its cultural evidence is clearly demonstrated by faience objects, copper beads, coins and red ware of late Harappan group. These remains cover an area of 8133 square metre and rises to a maximum height of one metre from the surrounding surface level in the west.

The site is situated on the dividing line of the two zones—the Bhangar and the Khadar. Since it is meeting point of elevated aeolian soil in the west with reasonable quality of soil water and alluvial low-lying fertile soil in the east with tremendous fresh water it could have offered opportunities in all the conditions both for agricultural and pastoral activities.

32. *Harsana Kalan – II: RW, Kushan and E. Med. 28° 57' N 77° 01' E*

This is another ancient settlement in the revenue limits of the same village and it is located only 300 metre west of the previous mound and a road leading to Harsana Khurd. Harsana Minor runs through the mound. Nowadays these remains are disturbed by intensive farming.

Like the previous site these remains are also perched on rising aeolian soil. Its cultural history is revealed by frequent occurrence of RW shapes of early historical period. The site covers an area of 27236 square metre. And the longer axis run north-south along the extension of the depression in the east. Since the Khadar area is not very far off from here the micro-environment is equally good and conducive.

33. *Harsana Kalan – III: RW, Kushan 28° 57' 02" N 77° 01' 06" E*

This ancient settlement is situated just east of the village on gradually rising clay soil. This is also disturbed by farmers, but the very occurrence of RW and pottery shapes of Kushan period helps to determine its cultural identity. The area of its circulation is roughly 32122 square metre, if we go by scatter of cultural artifacts and there is not any sign of cultural deposition. Regarding micro-environment we can simply equate it with that of the previous settlements.

34. Harsana Kalan – IV: RW and Med. 28° 57' 02" N 70° 00' 03" E

This is one more settlement in the revenue limits of the village and is situated to its south on the rising aeolian sandy soil, which is now under intensive cultivation. To the west of the mound there is a shallow depression with north-west to south-east orientation, which receives rain water during monsoon season. The nature of its cultural remains is quite obvious for the presence of RW shapes of early historical and medieval times. The extension of these remains is in north-west to south-east direction. And the geographical conditions are akin to that of the area.

35. Ibrahimpur Kurad: RW 27° 03' N 77° 03' E

The ancient settlement is situated one kilometre south-west of the village and across the canal flowing north-south. The ancient remains are located on rising aeolian sandy soil, which are in a disturbed condition for being under intensive cultivation. The nature of these remains is highlighted by occurrence of potsherds of early historical period. Its extension seems to have been north-south in agreement with the orientation of a shallow depression in the east. The nature of the subsoil water is good as it comes in the Khadar. And so is the nature of soil around it. These days the area is considered very productive.

36. Jakhauli: RW and E. Med 28° 56' N 77° 10' E

This ancient mound is situated in the east of the village at a distance of one kilometre. Locally this is known as 'Reda' — a mound. Its cultural identity is revealed by RW sherds of early historical and medieval times, which cover 22136 square metre area. The mound rises to 0.6 metre from the surface level. And the eastern parts tend to slope eastward in consonance with the general tendency of the plains. The site is surrounded by saline soil, but the subsoil water is quite good for being in the Yamuna Khadar.

37. Jawahari – I: PGW 29° 02' N 77° 01' E

The ancient settlement is popular by the name 'Kodala Kheda', which is situated one kilometre east of the village and just across the railway track from Delhi to Panipat. It stands west of the road from Sonipat to Bhurri and just north-west of Sonipat Sugar Mill. The present physical condition of the mound is not good simply because it is under intensive cultivation on the one hand and on the other experiencing attacks from the railwaymen and local people to obtain earth from here. However, the cultural association is quite obvious with the presence of PGW sherds and similarly the area of circulation, which is about 8073 square metre. Its orientation is almost east-west in consonance with the flow of natural depression to its north in close vicinity.

Regarding its surrounding micro-environment certain points are very interesting. First of all the location of the settlement on an elevated pocket just on the converging point of the Bhangar and the Khadar, which helps it to assume a special position of overlooking the low-lying Khadar in the north and the east. Secondly, the nature of soil in both the geographical zones—the Bhangar and the Khadar is loose because of dominant percentage of sand. Thirdly, subsoil water is fresh along the line of demarcation and inside the Khadar belt. And finally the uneven surface lends credence to absorb inherent shocks in the realm of climatic conditions.

38. Jawahari – II: RW and E. Med. 29° 02' 01" N 77° 00' 08" E

This is the second ancient settlement in the revenue record of this village and is situated 400 metre north-west of the foregoing mound. This was also perched atop a rising alluvial sandy pocket, which comes well inside the Khadar flood plain. At present the mound is under frequent cultivation and almost levelled by farmers to convert into an agricultural field and similarly, railway department has also collected earth to construct second railway line parallel to the first one. As a result only a scatter of RW pottery shapes belonging to early historical and early medieval periods is visible. It is difficult to find out its accurate size, but atleast it is sure that the site enjoyed a reasonable size before this damage. This is further supported by the version of village seniors. So far as the surrounding environment is concerned we can simply state that it is more or less akin to that of the former settlement.

39. Joshi Jat: RW and E. Med. 28° 58' N 77° 04' 05" E

This is situated on the southern periphery of the village. It is partly under present habitation and partly under cultivation. Hence, the present state of affair is not healthy except the fact that still enough potsherds of RW of early historical and early medieval periods are remaining to establish its cultural sequence. But so far as the systematic index of the cultural development is concerned there is nothing of the sort. Even precise or near accurate dimension of the site is difficult to realize. In such a situation we have to rely heavily on estimation/guesswork, which brings its approximate size to 38236 square metre.

Since the site is situated inside the Yamuna Khadar both the soil types and water are very conducive. Moreover to the south-west of the mound existence of a low-lying soil, which might have been of greater size in the past, is an added qualification. Soil water level is close to the surface and the soil enjoy appreciable water absorbing capacity. These things are very crucial in the life of any rural society.

40. Kakroi: RW, Kushan and E. Med. 28° 57' 40" N 76° 57' 30" E

This ancient settlement is situated 1.5 kilometre to the north-east of the village on aeolian sandy soil in the fields of Pala Ram Sarpanch, 5 kilometre south-west of Sonipat. Locally it is known as 'Khidkiyan' or 'Bhaiyan'[34]. The present condition of these remains is not intact because these are under intensive cultivation. But there are sufficient remains in the form of potsherds to establish its cultural identity with early historical and Sunga-Kushan period. And similarly, these very remains also help to estimating the size, which comes to 17800 square metre. This rises to 1.5 metre from the adjacent surface level. Their longer axis run north-west to south-east.

To make out the existing environment around the site we can take resort to soil types and nature of subsoil water. With regard to soil types we cannot escape the description of aeolian sandy soil around the mound, level sandy loam soils in the east and south and relatively low soil of clay loam nature in the south-west. Similarly, the nature of the original subsoil water is brakish to saline, again of regional character, in the deeper aquifers, whereas in upper stratas it is of working order, mainly effected by the canal system—the Western Yamuna Canal. The vegetation is of sparse and dry nature like the usual character of the Bhangar. Consequently in the summer season it is inhospitable.

41. Kami: RW and E. Med. 29° 03' N 77° 02' E

This ancient mound stands 0.5 kilometre west of the village

and 200 metre east of Karnal railway track and jutting on the dividing line of the Khadar and the Bhangar. This is placed on an alluvial mound, which is locally called 'Khera' by the people. Its present condition is not good, since disturbed by farmers and others. However, sprinkle of RW sherds belonging to early historical and medieval times speak of its cultural identity and sequence. It rises to 0.5 metre from the adjoining surface level and the longer axis are placed in the north-south position, perhaps dictated partly by the flow of the depression in the east and partly by the availability of suitable space for expansion. The approximate size of the mound is 32136 square metre as demonstrated by the spread of pottery. Since the settlement is inside the Khadar the natural surroundings are promising provided effective tools and techniques are available to harness to their benefit.

42. Kharkhauda : RW and E. Med. 28° 52' 05" N 76° 54' 02" E

The ancient mound is situated to the south of modern town of Kharkhauda, which is now occupied by huts and a building of cooperative society. Its elevation from surrounding surface is 1.2 metre. And only RW sherds were visible to mark its cultural background. It was difficult to ascertain its size in the given conditions.

In regard to micro-environment of the settlement we can mention aeolian sandy soil in the north-east behind Kharkhauda bus-stop, level sandy loam soil in the east and level clay soil in the south and west. A shallow depression exists in the west of the site and across the Kharkhauda-Matindu Road. Besides the nature of the sub-soil water is brackish to saline barring the shallow aquifers wherein water of working order is available in limited amount. This is induced by frequent visit of canal water over the years. But in the ancient times in absence of such a canal net-work people might have sought resort to the natural depression to arrange water. Hence, the situation is like that of the Bhangar character.

43. Kheri Tega: RW and E. Med. 29° 08'N 77° 08' 04" E

This ancient settlement is situated just east of the village and is now under intensive cultivation. It has been levelled by farmers, but the stray evidence of RW pottery shapes betrays its cultural identity. And if we go by the spread of pottery its size is 9133 square metre. There is no elevation, whatsoever, but the longer axis seem to have remained in the north-south position in accordance with the flow of Yamuna waters. Being very close to the Yamuna its surroundings are interesting including the constant flood threat from the Yamuna.

44. Kundal – I: RW and E. Med. 28° 50' 06" N 76° 57' 08" E

The ancient settlement is situated 0.5 kilometre north of the village and east of the Saidpur-Kundal Road[35]. This is placed on a natural mound of clay soil. From all the corners farmers have disturbed it. But still sizeable portion is intact, which is being spoiled by earth diggers. Its orientation is north-south and the height is 1.3 metre from surrounding surface level. Its covering area is 32136 square metre. Frequent occurrence of RW pottery shapes of early historical and medieval periods exhibit its cultural association.

The main soil types around the site are of the following order. In the west there is a stretch of aeolian sandy soil with the gentle rising tendency, which is afflicted with salinity in parts. To the south and east the soil is of clay loam type and tend to slope further south-east. In the north the soil is saline. The quality of the subsoil water is brackish to saline and the cover of vegetation is also accordingly sparse and alkaline. Most of the plants are salinity and drought resistant. All these do not inspire to draw a rosy picture. Despite these natural hardships people managed their existence, which is quite interesting in itself.

45. Kundal – II: RW and E. Med. 28° 50' 01" N 76° 57' 07" E

This is another ancient site in the revenue limits of the village and is occupied by present habitation, which is rising to several metre. At certain exposed points and in rain gullies RW pottery shapes of early historical period are met with to establish its identity[36]. Its dimensions could not be achieved because of present habitation. The macro-environment is more or less like that of the first settlement.

46. Laharada – I: LH 28° 58' 15" N 77° 00' 45" E

This ancient mound lies two kilometre north-west of the village on a rising sandy pocket known as 'Bheda-Thali'[37], which forms fields of Mange Ram and Sada Ram. At present the mound is under cultivation. Its cultural background is marked by occurrence of late Harappan sturdy red pottery and pieces of faience bangles. These remains are spread over an area of 9236 square metre and rise to 2 metre from the surrounding surface level.

The soil types are very interesting, since combination of aeolian sandy rising and sandy loam soils around the site is visible. The quality of soil water is brackish and therefore, in absence of any permanent source of fresh water in the close vicinity of the settlement people managed their existence is interesting. If we walk down 2.5 kilometre in the east a sharp depression is met with lying east of the village and Sonipat-Kharkhauda Road, which in fact is the extension of the Yamuna Khadar. This contains fertile alluvial soil and fresh water. In a way the ancient site is placed on the rising western line to enable the people make use of the two geographical zones according to changing climatic conditions as we witness nowadays.

47. Laharada – II: GW, RW and E. Med. 28° 58' 15" N 77° 00' 45" E

This is another ancient mound in the revenue limits of the village and is located 1.3 kilometre north-west of the village[38] and just 300 metre east of the previous settlement. Near to this is situated Hafed godown. It is under cultivation, but potsherds belonging to GW and RW of early historical and medieval periods are freely available. It seems to have been of considerable size as suggested by spread of pottery. However, precise dimensions are difficult to manage in the given situation. With regard to its micro-environment we have already mentioned while dealing with the first settlement.

48. Laharada – III: RW and Med. 28° 58' 15" N 77° 00' 15" E

This is one more ancient settlement in the area of the village and is situated 1.5 kilometre west of the village on an aeolian sandy pocket, which is known as 'Thali' locally[39]. The site now forms fields of the Bharat Singh, Ram Singh and Rai Singh. At present it is under intensive cultivation and therefore, is in a disturbed state. Scattered evidences of early historical RW are indicative of its cultural association. The present elevation from the surface level is 3 metre and the encompassed area is 38136 square metre. The longer axis are north-south. And the surrounding environment is akin to those of the previous two settlements in the same area.

49. Mandori: RW and E. Med. 28° 52' 05" N 76° 59' 08" E

The ancient remains are located to the south of the village among elevated clayey fields. At present these are badly disturbed for their being under intensive cultivation. The early historical and medieval ceramic wares speak of its cultural background. These remains are found over an area of 22421 square metre. The natural environments are not favourable in any manner.

50. Mohana : RW and E. Med. 29° 02' N 76° 52' E

The ancient site is situated one kilometre west of the village on the left-hand side of Sonipat-Gohana Road under Haryana Government High School building. It is spread over to 24563 square metre and contains RW pottery types of early historical period to mark cultural identity. The soil around the site is sandy loam and suffers from increasing salinity. Water quality is of working order, but only in shallow aquifers. Towards south the surface level is receding increasingly to join the depression, wherein rain-water accumulate in rainy season.

51. Moi: RW and E Med. 29° 02' N 76° 43' E

This ancient site is situated one kilometre south of village Moi on a *kuccha* road. It is almost plain being under cultivation. Its cultural identity is betrayed by scatter of RW shapes, terracotta beads and other objects. Similarly pottery types of early medieval period were also collected. These remains cover an area of 42134 square metre.

The geographical scenario does not look very conducive despite the fact that the nature of sub-soil water in the upper strata is of working order. However, the soil is ridden with encroaching salinity around the settlement, though, of course, the nature of the soil appears to be sandy loam. During the summer season the emerging picture appears very dismal, which begins to turn with the change in season from summer to monsoon. This converts it into a thriving green land. So it is essential to bear in mind all such micro-ecological aspects while dealing with ancient societies.

52. Murthal: RW, Kushan and E. Med. 29° 02' N 77° 05' 04" E

This ancient settlement is situated on the north-eastern fringe of the village and on the left-hand side of the road on an elevated alluvial soil. This is completely under intensive farming and the remains embrace a vast area of 63236 square metre with 1.2 metre elevation. The longer axis are in the north-south position. RW and Kushan pottery shapes are littered over the entire landscape to point out its cultural history.

Its surroundings are quite interesting for the site is well inside the Yamuna Khadar where fertile alluvial soil with tremendous amount of fresh water is not scarce. Moreover in the south exists a huge north-south running depression, locally known as 'Dahar' or 'Bahaya', usually full of water, is an added positive feature. These days the area is providing tremendous opportunities to the people for their subsistence.

53. Nahari – I: RW, Kushan and E. Med. 28° 52' N 77° 02' 07" E

This is situated on the south-western fringe of the village and north of Narela-Nahara Road among level agricultural fields. Since the mound is completely under cultivation and levelled by farmers it is difficult to mark out its exact size and height, though it is said to have been a big mound in the past rising above one metre as pointed out by the villagers. However, under these circumstances it was equally problematic to identify its cultural association, but for presence of RW sherds of early historical and medieval periods.

As regards micro-environment we can state its location on the dividing line of the Bhangar and the Khadar. The former enjoys saline soil water and high clay or sandy loam soils, whereas the latter, though low-lying and form part of the Khadar, yet, lacks in typical Khadar characters. Here neither the soil water is good nor the soil is primarily alluvial in character. However, the depression locally known as 'Dahar' retains water for greater part of the year in normal rainfall conditions and the mixed soil is also capable of producing bumper crops. To the north of this site there is a vast stretch of rising aeolian sandy soils. Hence, the combination of these factors make it an interesting environment.

54. Nahari – II: LH, RW, Kushan and E. Med. 28° 53' N 77° 20' 58" E

This is another ancient settlement in the revenue record of the village and is situated 1.3 kilometre north of the village on rising sandy pocket. This is also under intensive agriculture and as a result it is difficult to manage its exact dimension and orientation, yet, the approximate size is 1990 square metre. However, if we go by the spread of pottery and the orientation of depression their extension would be in the north-south position. These remains highlight its cultural background also. The occurrence of late Harappan sturdy pottery and faience bangles in the west speak of its first ever occupation sometime in this phase. Similarly greater frequency of RW pottery types of early historical, Kushan and medieval times over the length and breadth of this settlement underlines its subsequent possession by these people. Since the site is situated in the same locale the micro-environment is akin to that of mentioned before.

55. Nakloi – I: PGW–BSW 28° 58' N 76° 52' E

The ancient site containing PGW and BSW is situated 400 metre east of the village. And is locally known as 'Gujarali-Khera'. At present the mound is under intensive cultivation and completely disturbed, which make it difficult to assess its precise expansion, but earlier its size was 5 acres as suggested by Silak Ram[40]. However, the spread of pottery amply demonstrates its north-west to south-east orientation.

The natural ambience is not conducive in absence of any permanent source of water either below or above the surface. The soil is also not usually good as it suffers from increasing salinity and supports poor alkaline vegetation. And if we presume to take away existing facility of modern times the area would prove very hostile.

56. Nakloi – II : GW and RW 28° 58' N 76° 52' E

This is another ancient settlement in the area of same village and is situated only 300 metre north-west of the village. It is also in a dilapidated condition, but presence of GW and RW potsherds indicates its cultural association, though these are not very inspiring. The spread of these remains is marked on an area of 4020 square metre, but previously this was 4800 square metre and 1.85 metre high[41]. The environment is akin to that of the previous settlement.

57. Nangal Khurd: GW and RW 29° 02' N 77° 05' E

The ancient site is located east of Grand Trunk Road and south of the approach road to the village. This is locally known as 'Bania-Wali'. At present the mound is completely disturbed by earth diggers and the remains are very scanty. Only a few sherds of GW and RW of early historical period were collected to verify its cultural identity. Since the site is situated in the heart of Khadar its environment is not disappointing. In addition to the east of the site exists a north-south running natural depression known as 'Dahar' or 'Bahya'. This might have served as a source of surface water in the past.

58. Panchi Gujjran: GW and RW 29° 11' N 77° 02' E

The ancient settlement is situated 2 kilometre west of the village and one kilometre east of Garhi Jhajjara village[42]. The site is surrounded by green fields and at present it is almost intact except peripheral areas where farming is practised. This is a huge mound covering 36136 square metre area of alluvial sand and rises to 2.8 metre from the surrounding surface level. Its cultural antiquity is attested by the presence of GW sherds spread in the north-south position. In addition pottery of early historical period speaks of its subsequent occupation. Since the surrounding area form part of the Khadar the area is conducive for all purposes.

59. Pipli Khera: GW-BSW and RW 29° 07' N 76° 04' E

The ancient mound lies 300 metre west of the village[43] and just east of the canal. This is locally known as 'Khera'. The earliest most cultural remains belong to PGW levels. Such remains were available only in the south-western section where the portion is exposed because of digging by the villagers and erosion by rain-water. In this section only some potsherds of BSW were also picked up to suggest presence of such remains along with GW. Similarly circulation of RW pottery of early historical period in the remaining parts denote its subsequent occupation in this period. In regard to its habitational expansion what we observe is north-west to south-east orientation and its area is 12123 square metre with less than 3 metre height.

Topographically the site is placed on elevated alluvial pocket with very fertile arable land around enjoying fresh subsoil water. The surface level tend to slope down gently to the Yamuna in the east. A shallow depression running in the same fashion lies to the north of the mound. Similarly in the south the area is low-lying and seems to have experienced water conditions for a considerable time period.

60. Pipli – I: LH-OCP 28° 51' 06" N 76° 56' E

The ancient remains are lying atop a sandy mound at a distance of 800 metre from the village in the south.

This is locally known as 'Thali' and people also call it 'Chhota Khera'. This is now completely under constant cultivation and the remains are in a disturbed state. In the given situation neither it is possible to judge patterns of expansion nor can we assess cultural thickness. But potsherds of OCP and late Harappan group along with pieces of faience help establish its cultural antiquity. The area of circulation of these remains is very small covering only 2040 square metre.

Its surroundings present a significant combination of various types of soils. The aeolian sandy pocket is stretched in the north-south direction with varying elevation and dimension. This is followed by level sandy loam soil in the south and the east. But to its north-west there exists one natural depression, wherein nature of the soil is clayey. The nature of soil water is fresh, but that is restricted to the upper levels only and below that it is of saline nature. Since the settlement is situated in the Bhangar the average vegetation is sparse and primarily consisting of drought and salinity resistant species. However, the general appearance is not disappointing.

61. Pipli – II: RW, Kushan and E. Med. 28° 51' 05" N 76° 56' 01" E

This is another ancient settlement in the revenue limits of the village and is situated 400 metre south--east of the previous settlement on the easternmost extension of the sandy ridge, just west of Rampur Minor. These days the site is under frequent cultivation and is levelled by farmers. The very spread of early historical, Kushan and early medieval RW shapes brings out its expansion on the one hand and on the other highlights its historical background. And similarly east-west extension pattern is demonstrated by the spread of ceramic wares. As regards micro-environment existence of low-lying area in the south is noteworthy. Other features are like that of the previous site.

62. Qamaspur : GW-BSW and RW-NBPW 29° 00' N 77° 06' E

This mound is one kilometre east of the village[44] and at equidistance from the Grand Trunk Road in the west. The mound is partly disturbed by farmers and partly by villagers in order to collect sand from here. In a way the overall condition is not good. Its cultural history begins with GW remains and followed by RW-NBPW. These remains are spread over an area measuring 12137 square metre and rises to 1.6 metre from adjoining surface level. The cuttings in the northern and eastern sections reveal a thick deposition of alluvial sand meaning thereby this occurred before the mound was occupied sometime in the GW levels. This is very significant to throw some welcome light on the behaviour of the Yamuna. The environment is good for its location in the Khadar. But surprisingly there exists a vast stretch of saline soil (contrary to the general nature of the Khadar) to the west and south of the mound. However, the nature of the soil water is favourable.

63. Rewali: GW and RW–NBPW 29° 01' N 77° 04' E

This ancient settlement is located outside the village[45] in the north and to the left-hand side of the road from Sonipat to Murthal. At present it is under cultivation and the physical condition is not good. Its cultural history seems to begin with GW remains and subsequently followed by pottery shapes of early historical and early medieval periods. It is difficult to recover length and breadth of the settlement and therefore, it was estimated to 38233 square metre without any elevation.

The mound fall within the Khadar limits, especially in the central part. The surface level tend to slope in the south-east direction, i.e., to the Yamuna river. There exists narrow, long and shallow depression, north-south oriented, which tends to embrace Drain No. 6 from Sonipat. Locally people call it either 'Dahar' or 'Bahya', which runs west of the mound. Natural conditions are favourable for dry farming as well as wet cultivation.

64. Rolad Kalan: RW and E. Med. 29° 06' 04" N 76° 51' E

This ancient settlement stands in between Rolad Kalan and Rolad Khurd villages on an elevated sandy mound. These days the site is in a disturbed condition caused by multifarious activities of the villagers. The approximate size of the mound,is 36536 square metre, which rises to 3.2 metre with north-west to south-east expansion pattern.

A large number of RW pottery shapes of early historical and medieval times underlines its cultural antiquity. Since the site is in the Bhangar the natural conditions are not very promising except presence of sandy pockets in the south and low-lying area in the north and north-west, which receives rain-water from Bhadi in the west and advances to Dodwa in the south-east.

65. Sandhal Khurd: GW and RW 29° 03' 20" N 76° 58' E

The ancient site lies to the north of the village at a distance of 300 metre on the rising conjunction of the Bhangar and the Khadar. The site in fact lies in the revenue record of this village instead of Sandhal Kalan as reported by Phogat[46]. The road from Sonipat to Purkhas runs through dividing it into two halves. This is overlooking the Khadar in the east and now is under intensive cultivation. It embraces an area measuring 8136 square metre as suggested by the spread of pottery.

Regarding micro-environment we can mention incoming depression forming a dividing line, the Western Margins, of the Bhangar and the Khadar. To its east the soil is consisting of alluvial sand enjoying fresh water while in the west it is sandy loam with brackish water. Thus, two contrasting zones are promising a wide range of avenues.

66. Sandhal Kalan: RW and E. Med. 29° 08' N 76° 14' E

This ancient settlement stands on the north of Sandhal Kalan road leading to railway station of the same name. People generally call it 'Bada-Johad', since the mound is situated near the big village pond. It overlooks the incoming depression in the east. The present condition of the site is not good. Potsherds of early historical and medieval times clearly demonstrate its antiquity and cultural sequence. Its micro-environment is similar to that of the previous site, since these two are located in uniform natural environment.

67. Sewali: RW and E. Med. 28° 56' N 77° 11' E

This ancient mound lies to the east of the road from Sewali

to Jakhauli on an alluvial rising pocket, which is now partly disturbed by some cuttings in the south-eastern parts. This is a vast mound covering 54136 square metre area with east-west axis. Red ware pottery shapes of early historical and medieval periods are freely available to establish its antiquity.

The area around the site is having a gentle slope to the south-east, where Drain No. 8 runs to join the river Yamuna in the east. Since the area is a part of the Khadar the soil is alluvial and very fertile. Subsoil water is fresh. The natural features are very interesting. The Yamuna is flowing 7 kilometre in the east.

68. Shahazadpur: RW and E. Med. 29° 02' 05" N 76° 59' E

This ancient mound is situated 400 metre south-west of the village and the road from Sonipat to Purkhas. Now it is covered by Haryana Government School. Thus, the site is disturbed and poses problem in taking its precise size. There is hardly any elevation these days. But despite all these problems its antiquity can very well be established with the help of available RW potsherds of early historical and medieval periods. Besides some copper coins are said to have been collected by Anand Swarup. In order to have a look, if not to collect, on the coins I visited his house twice but could not realize this hope. Anyway this is also an interesting site on the dividing line, the Western Margins, of the Bhangar in the south and the Khadar in the north. Therefore, the existing environment is quite significant.

69. Silana: EH 28° 55' 09" N 76° 50' E

This ancient settlement is situated north of the village on an aeolian sandy pocket among fields[47]. Locally people call it 'Bade-Wala'. Being under intensive cultivation its size and cultural deposition are difficult to manage. But spread of early Harappan pottery of thin and red section, plain surface and evolving rims speak of its early Harappan cultural association. Its surroundings are not promising in any manner as bear similarity with the broad spectrum of the Bhangar zone.

70. Sisana – I: LH and GW 28° 53' N 76° 53' 02" E

This ancient site is located 3.5 kilometre east of the village and on the north of Kharkhauda-Rohtak Road[48]. These days the site is under intensive cultivation and cultural deposits have been levelled by farmers, but some potsherds of late Harappan and

GW along with pieces of faience bangles and beads are quite obvious to indicate its cultural identity. The area containing these remains is 990 square metre only. Natural depression in the north through, which Drain No. 8 runs is quite significant to point to its micro-environment.

71. Sisana – II: LH and PGW 28° 52' N 76° 50' 07" E

This is another ancient settlement in the revenue limits of this village and situated 3.2 kilometre south of the village on a rising loam soil. It is now levelled by farmers and the cultural remains have been dumped on the local passage. Here late Harappan pottery along with pieces of faience bangles and PGW sherds are frequently not only noticed to establish cultural identity and sequence, but also its size of 4020 square metre area. But these days the surroundings are not conducive in any manner because the soil is replete with salinity and brackish water. A shallow depression lies in the south, which is occupied by a local drain with north-west to south-east orientation. In the north-east, near the Gurukul of Matindu village, a sandy pocket is also significant.

72. Sonipat: PGW–GW and NBPW 29° 00' N 77° 01' 40" E

The ancient settlement stands west of Sonipat Main Bus Stand and under almost complete occupation of modern habitation. This is, of course, quite distinct for its height and huge size. Remains belonging to the aforesaid cultures were collected, though their number is small, from some exposed portions and sections. Yaudheya copper coins have also been mentioned previously from this settlement[49]. The location of this site is quite interesting because on the one hand it is placed on the junction of the Bhangar and the Khadar and on the other on an elevated mound mainly composed of alluvial sand. Therefore, it overlooks the low-lying Khadar in the east. To the east of the mound exists a deep natural depression in the north-south position, which is now obliterated by modern development works. Drain No. 6 runs almost through this depression as an interminable testimony. Moreover in this section thick concentration of the Yamuna sand is clearly visible on the surface where nature of soil water is also extremely good. All these features can very well be identified while traversing through Haryana Housing Development Board Sectors 14 and 15, located just east of the mound. If we have a look around in the west none of these features is encountered with for nature of soil is largely sandy loam with deep-seated water-table of hard nature. Corresponding difference in the nature of vegetation can easily be marked. Thus, the settlement is standing at a very crucial topographical point to enjoy natural qualities of these two geographical distinct zones.

73. Tajpur: RW and E. Med. 29° 04' N 77° 08' E

This ancient mound is situated south-east of the village at 0.5 kilometre on a rising pocket mainly composed of alluvial sand of the Yamuna river, which now mainly forms premises of Government High School. Though the ancient deposition is not intact, yet, it is not difficult to identify its cultural record mainly because presence of pottery remains of early historical and medieval times. The system of expansion seems to have been in the north to south direction in accordance with the flow of the Yamuna. Its approximate size is 36327 square metre without any cultural depositional elevation. Since the site is in the Khadar, the surrounding environment is quite promising. In addition in the west of the site there exists a local depression with north-south orientation, which was full of water, when the area was visited by me. But the level of subsoil water is 6 to 8 metre deep, which is not in agreement with the Khadar qualities. This seems to have been a creation of modern agricultural and ecological conditions. As regards the former a large network of tube-wells exploit the sub-soil water intensively for paddy and wheat crops, which is occasionally aided by negative fluctuation in rainfall pattern reducing further the rate of refreshment of underground reservoir. The process of recharging is further hampered by the recently formed earthen bund all along the western bank of the river Yamuna designed to protect people from unpredictable flood havocs. With its appearance the process of annual inundation is precluded,with the result the inherent refreshing mechanism is further blocked. Thus, the whole scheme calls for redesigning in the light of the above facts to make it a success, environment friendly and in the long run interest of the society.

74. Thana Khurd: RW and E. Med. 28° 54' N 76° 57' 09" E

The ancient settlement is situated one kilometre west of the village on a sandy mound in the north of the canal. Locally this is known as 'Asthal' or 'Thali'. Being under cultivation cultural remains are in a disturbed condition. To identify its cultural background a sprinkle of pottery shapes of early historical and medieval times is sufficiently available. Keeping in view the spread of potsherds we can assess its probable expansion and hence, it roughly covers 27136 square metre area. Its longer axis seem to exist in north-west to south-east position according to the orientation of a natural depression in the east.

With regard to the extant surrounding ambience we can take into account aeolian sandy, sandy loam and clayey soils in and around, fresh water in shallow aquifers, the sharp depression in the east and the alkaline nature of general vegetation. These features offer a number of avenues for subsistence activities.

75. Zhinjhauli: RW 28° 51' N 77° 00' 01" E

This ancient mound is situated to the south-west of the village on sandy loam soil. Locally people call it 'Reda', which is now under intensive cultivation. The present physical state of this mound is not intact like many others in the area. But its association with the ancient past is quite obvious by the sprinkle of potsherds. These very potsherds are consisting of RW pottery shapes of early historical period. Therefore, it seems to have been a single-culture settlement. So far as other details of its dimension and expansion patterns are concerned there is hardly any concrete information for the disturbed state of affairs.

In connection with the micro-environment we can state the following points. To the east of the site exists a vast sandy ridge extending north-south. In other directions the nature of the soil is sandy loam with undulating surface. The quality of water is good, but limited to the upper levels only and that too in areas, where canal water is used for irrigation for quite sometime now. The nature of the water in deeper aquifers is again saline (of the Bhangar character). Similarly the original vegetation cover is in agreement with the prevailing conditions. Such a situation is promising a wide range of subsistence opportunities with the change in seasonal conditions as we experience these days.

Conclusion

The foregoing description of the explored settlements is quite interesting. It becomes obvious that the area under discussion experienced feel of human habitation sometime in the early

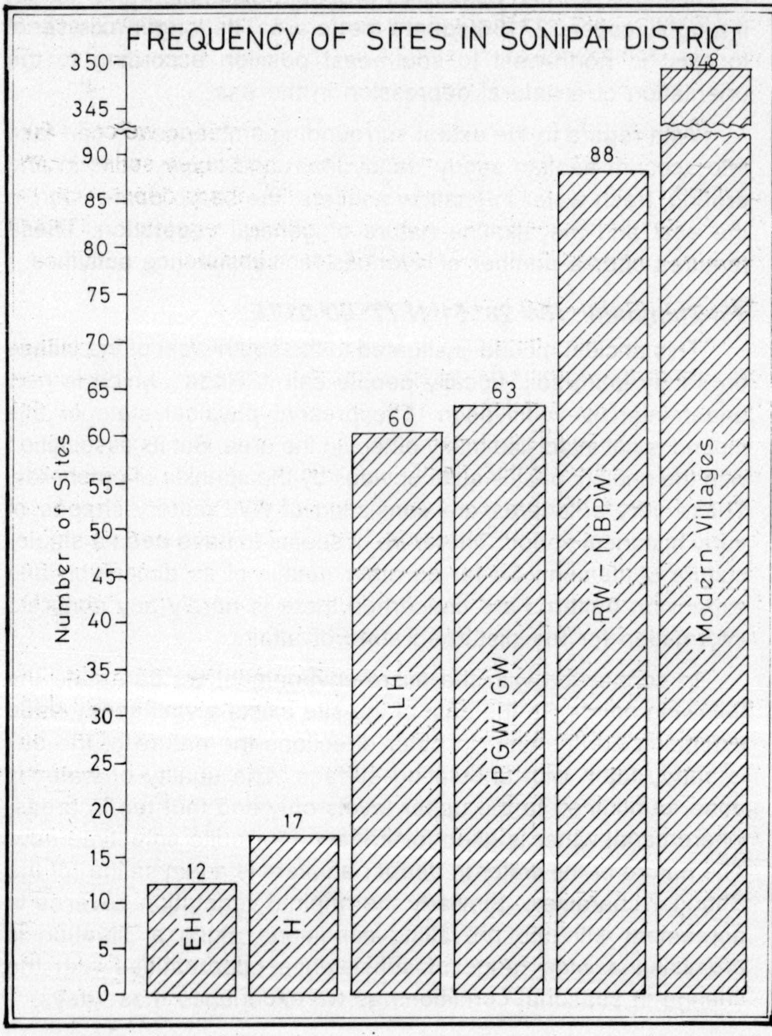
FREQUENCY OF SITES IN SONIPAT DISTRICT
Number of Sites
350
345
90
85
80
75
70
65
60
55
50
45
40
35
30
25
20
15
10
5
0
12
EH
17
H
60
LH
63
PGW-GW
88
RW-NBPW
348
Modzrn Villages

Fig. 2.1

Harappan times with the emergence of settlements belonging to this phase. And with the beginning of human life the area seems to have remained under occupation at various points of time till now irrespective of the obvious intervening gaps (Fig. 2.1). And as time went on we observe an increase not only in the number of settlements, their size and area of influence, but also improvement in the material culture. Similarly over the period of time a gradual evolution process is indicated in their techniques of subsistence, though with certain retrograde changes in between also. In the same way the increasing size of settlements points towards agradational patterns in the realm of population density at individual settlements and in context of different geographical locales and point of time. All these aspects might have been influenced by the prevailing micro-environment and social conditions. Therefore, there is a need to assess the past societies in the light of above factors, which will be taken up in the following chapters.

NOTES AND REFERENCES

1. Phogat, S.R., *Archaeology of Rohtak and Hisar Districts of Haryana*, Ph.D., Thesis (unpublished), Kurukshetra University, Kurukshetra, 1972, p. 35.
2. *Ibid.*, p. 32.
3. *Ibid.*
4. *Bhan*, S., *Mitathal Excavations and Other Exploration in the Sutlej-Yamuna Divide*, Kurukshetra University, Kurukshetra 1975, p. 125; Lal, B.B. and S.P. Gupta, eds. *Frontiers of Indus Civilization*, Delhi 1984, p. 525.
5. Phogat, 1972, p. 33.
6. *Ibid.*
7. Kumar, Ashok, *History and Archaeology of Kathura Block, District Rohtak-Haryana,* M.Phil Dissertaion (unpublished), Maharshi Dayanand University Rohtak, 1989-90, pp. 8-80.
8. Phogat, 1972, p. 34.
9. *Ibid.*
10. *Ibid.*, pp. 34-35.
11. *IAR*, 1954-55, p. 66.
12. Phogat, 1972, p. 34.

13. *Ibid.*
14. *Ibid.*, p. 36.
15. *Ibid.*, p. 37.
16. *Ibid.*
17. *Ibid.*
18. *Ibid.*, p. 38.
19. Bhan, 1975, p. 125.
20. Phogat, 1972, p. 38.
21. *Ibid.*, p. 56.
22. *Ibid.*, Bhan, 1975, p. 126; *IAR*, 1960-61, p. 65.
23. Singh, S., *Archaeology of Sonipat*, M.Phil. Dissertation (unpublished), Kurukshetra University, Kurukshetra 1984, p. 16.
24. Bhan, 1975, p. 125.
25. *Ibid.*, p. 58.
26. Dutt, 1982, p. 43.
27. Bhan, 1975, p. 125; *IAR,* 1960-61, p. 65.
28. Phogat, 1972, pp. 58-59.
29. Dutt, 1982, p. 44.
30. *Ibid.*, p. 59.
31. *Ibid.*, p. 39; Bhan, 1975, p. 125.
32. *Ibid.*, p. 59; *IAR,* 1960-61, p. 65.
33. *Ibid.*, p. 60.
34. Singh, 1984, pp. 17-18.
35. Phogat, 1972, p. 61.
36. *Ibid.*
37. Singh, 1984, pp. 16-18.
38. *Ibid.*
39. *Ibid.*
40. Phogat, 1972, p. 61.
41. *Ibid.*
42. *Ibid.*, pp. 61-62.
43. *Ibid.*, p. 62.
44. *Ibid.*, p. 60

45. *Ibid.*, p. 62.
46. *Ibid.*, p. 63.
47. *Ibid.*, p. 55.
48. *Ibid.*
49. *Ibid.*, p. 63; *AI*, 10 and 11, New Delhi, p. 145.

3

DISTRIBUTION OF SETTLEMENTS

The district of Sonipat was earlier a constituent of district Rohtak as has been said in Chapter I. Since it formed part of the said district it is imperative to take into account the past studies on these areas, especially relating to settlement geography. Such an approach to Indian archaeology is interesting and gradually gaining currency. The area under discussion is lacking in archaeological studies in general and settlement studies in particular for a variety of reasons. However, to be specific in the present context we can not, but refer to Dutt's work in brief as it has already been discussed in detail in Chapter I. This deals with the problem of settlement geography of PGW in Haryana State. As it is confined to a particular cultural period in Haryana, our area of discussion was bound to receive a general attention while dealing with settlements of the said cultural period[1]. And more so, since the work is primarily based on the already reported settlement data, especially in regard to our own area, by Phogat[2]. Nevertheless this work is not only of pioneering nature, but also offers an interesting insight into the sequence of spread and development of settlement hierarchy. The aforesaid study does provide a broad framework of distributional pattern to form an anchor sheet for any forthcoming intensive and analytical study. Thus, in the light of this, the present exercise assumes added significance. In this background the present exercise was undertaken definitely with a different approach and objective. And for that matter the intensive field-work was undertaken based on village to village exploration produced not only a substantial fresh data in terms of number of settlements to be added to the existing one, but also in terms of soil types, water resources, location of settlements and existing patterns of land use and crop rotation. The information thus, gathered constitutes an impressive body of protohistoric settlements and present subsistence patterns. All these imbibe a series of queries and their explanation regarding the past patterning of settlement geography and the mode of subsistence. Thus, on these lines an attempt to unfold the possible trends of the ancient past would be made in the following discussion.

Culture-wise Breakdown of Sites

The aforesaid exploratory exercise produced 189 settlements found scattered over the landscape. These present a wide range of cultural spectrum ranging from early Harappan to late medieval period. A little over half of the explored settlements belong to either the early historical or the different stages of medieval period. The remains of protohistoric cultures are associated with 92 settlements. The precise number of protohistoric and early historical sites is 145. And since these remains are central point to the present discussion, our immediate focus revolves around them. It is, therefore, appropriate to analyze the available data on these cultural lines in the following paragraphs.

On close examination it was observed that of the 145 protohistoric and early historical settlements only 12 belong to the early Harappan and early Siswal B[3]. The remains from these sites demonstrate a family likeness with the early Siswal culture. The succeeding mature Harappan is represented by 17 sites. These settlements bear cultural remains similar to those of Mitathal-II-A[4] and are equated with Siswal, which in the broader framework corresponds with mature Harappan[5]. This period is followed by Mitathal II-B[6], which is termed as late Siswal[7]. The cultural artifacts of this phase have been gathered from 60 settlements, an appreciable number in relation to the previous levels. This is succeeded by PGW and GW phase representing 39 and 24 settlements respectively. 88 settlements produced remains of RW-NBPW cultural assemblage in succession to PGW-GW[8]. This cultural division of sites (table 3.1) reveals the following picture.

TABLE 3.1

Sl. No.	*Culture*					*No. of Sites*
1.	Early Siswal-B	=	Mitathal-I	=	Early Harappa	12
2.	Siswal	=	Mitathal-II A	=	Harappa	17
3.	Late Siswal	=	Mitathal-II B	=	Late Harappa	60
4.	PGW-GW					63
5.	RW-NBPW			=	Early Historical	88

The cultural sequence exhibits a very significant point regarding inception of human life in this area. This began sometime in the second phase of the early Harappan period — early Siswal-B — a transitional phase between early Siswal-A and Siswal/Harappan. This shows a gradual dissemination of human settlements from the west to the east. Because in this zone signs of Kalibangan-I or early Siswal-A are conspicuous by their absence. And the cultural artifacts from here bear close affinities with those of the upper levels of Kalibangan-I, which is contemporary to the early Siswal-B. Thus, the phenomenon in this belt appears belong to later period than the areas in the west. And more so, if we take into account the uninterrupted cultural depositional process as recorded at sites like Rakhigarhi, Mitathal[9], Balu[10] and Banawali[11], in relation with the entire Harappan period (the various stages of Harappa civilization) then not only we can support our earlier observation, but also infer habitational continuity in regard to our own area, which, of course, remains subject to verification through future diggings. Likewise the overlap between late Harappan and PGW as discovered at Bhagwanpura[12] help assume similar conditions with regard to continuity in habitation from late Harappan to PGW phase in our own area, at least untill it is further confirmed either way through future diggings we can accept it for the time being keeping in view the geographical overlap between the two. This also highlights the terminal stage of late Harappan and beginning of the PGW. And similarly in order to trace the last stage of the PGW we can take help of other facts of similar nature from the surrounding areas. We may state chronological position of Sugh-IB[13], Raja Karan Ka Qila-I[14] and Autha-II[15], which form the uppermost limit of the PGW in Haryana. In these levels the PGW is limited and coarser. The frequency of GW is growing considerably[16]. And

PGW sherds at Sugh occur only in period-IA along with NBPW while the succeeding-IB yields only the GW[17]. This is evident that the PGW tradition in Haryana (as elsewhere in India) survived only in the early phase of NBPW, period. Hence, IB (early phase of NBPW) at Sugh marks the upper time limit for the PGW culture and therefore, comes in between the late Harappan and the early NBPW-Sugh I-B phase[18]. This proposition can further be verified only through subsequent excavations. Such a process of expansion is termed as 'wave of advance' or 'population expansion'[19]. But in our own case this model does not seem applicable, since we fail to perceive any wave as such. And what we witness is an indication of a gradual growth of population and dispersal of settlements till we embrace the late Harappan phase, wherein an upward upsurge is quite obvious in relation to average number of settlements and overall population size. This seems to be the result of various factors. Therefore, this is called a 'gradual advance dispersal' inspired, though partly, by increasing tendency in population. In the absence of any chronometric dates it is difficult to presume and assign the exact dates to these periods. However, in order to trace the chronological sequence, in terms of time period, of these cultures we can atleast make an attempt to assess their tentative time-schedule with the help of available C-14 and relative dates of these cultures (table 3.2).

TABLE 3.2

Sl. No.	*Culture*			*Period*	*Occurrence*
1.	Early Siswal-A	=	Kalibangan-I	C. 2350-2100 B.C.	No
2.	Early Siswal-B	=	Mitathal-I	C. 2100-2000 B.C.[20]	Yes
3.	Siswal or Mitathal-II A or Banawali-III	=	Harappa	C. 2000-1800 B.C.	Yes
4.	Late-Siswal or Late Harappa	=	Mitathal-II B	C. 1800-1200 B.C.	Yes
5.	Overlap Between Late Harappa and PGW phase			C. 1200-1100 B.C.	Yes
6.	PGW-GW			C. 1100-500 B.C.	Yes
7.	RW-NBPW			C. 600-200 B.C.	Yes

The chronological sequence given above supports our previous observation regarding beginning of human activities only in the second half of the early-Harappan phase and the area remained occupied in the succeeding periods. Let us now discuss the distribution of these settlements.

Distribution of Settlements

We have already noticed in the preceding discussion that 12 settlements produced remains of early Harappan culture. Of these 9 sites are located in the zone lying west of Gohana town (map 5). And that too in the south-western pocket of this zone. The main concentration revolves around Rindhana-Ghadwal belt. The locational analysis further shows that the northern parts of this western zone are devoid of any such settlement and that way it remained barren. The remaining 3 sites of this culture are of scattered nature as pointed out by their respective location at Farmana, Bahgru and Silana[21] (map 5). These hardly indicate any pattern of expansion though these signify that by this time people of this stock had somehow managed to penetrate hitherto in the east by way of 'gradual advance dispersal'. If these sites are viewed in the light of the available information on early Harappan settlements, one can easily establish a relation between them. There are sites like Bali[22] and Nilothi[23] in eastern parts of district Rohtak. These sites are close and in line with the sites we are discussing about. However, the central point of this period seems to have remained Rindhana-Ghadwal village complex, where seven of the total 12 settlements are located.

The succeeding levels witnessed existence of mature Harappan settlements. Only seventeen sites contain remains of this phase. The numerical strength of these settlements registers a marginal improvement over the previous phase. However, its functional significance is recognizable in terms of highlighting and reinforcing the continuation of the process of 'gradual eastward expansion', though still most of the sites are confined to the same zone, i.e., west of Gohana town. It further reaffirms the tendency to concentrate around the Rindhana-Ghadwal complex. Nevertheless in this phase the general pattern of distribution except the nucleus, is relatively even as sites are located at regular intervals (map 6). This distribution underlines once again, like the previous phase that the northern fringes of this zone still could not find favour

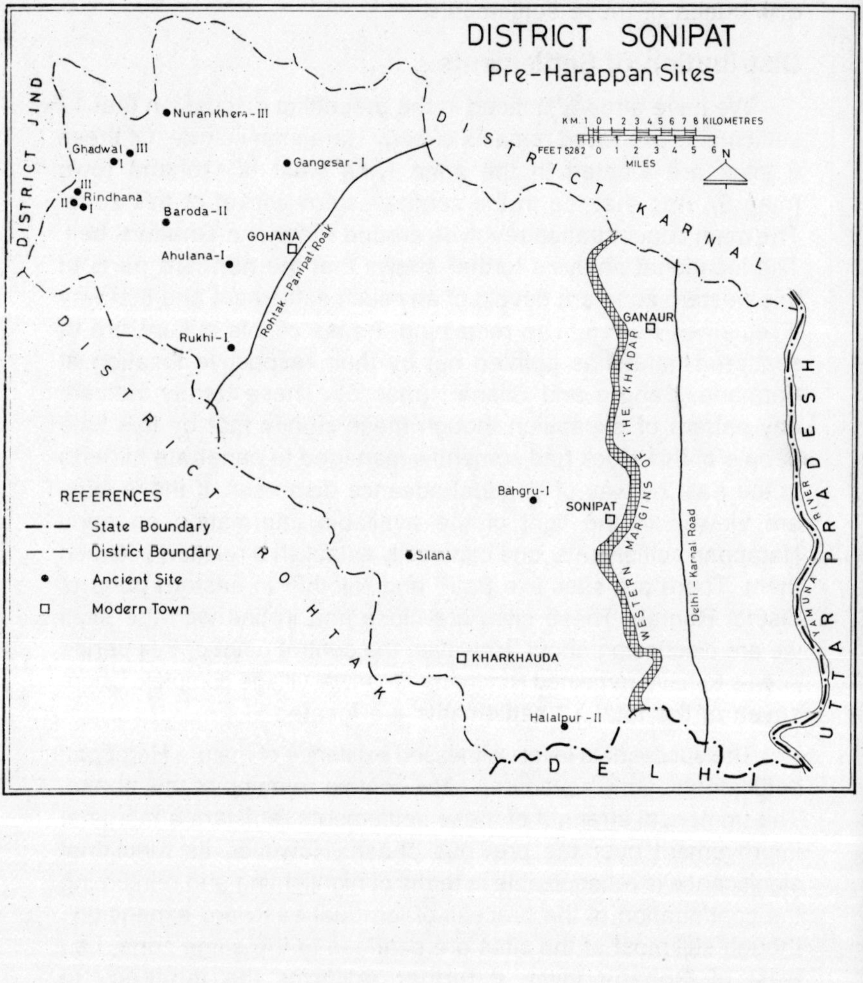

DISTRICT SONIPAT
Pre-Harappan Sites
KM 1 0 1 2 3 4 5 6 7 8 KILOMETRES
FEET 5282 0 1 2 3 4 5 6
MILES
N
DISTRICT JIND
DISTRICT KARNAL
DISTRICT ROHTAK
DELHI
UTTAR PRADESH
YAMUNA RIVER
THE KHADAR
WESTERN MARGINS OF
Delhi - Karnal Road
Rohtak - Panipat Roak
Nuran Khera - III
Ghadwal III
I
Gangesar - I
III
Rindhana
II I
Baroda - II
GOHANA
Ahulana - I
GANAUR
Rukhi - I
Bahgru - I
SONIPAT
Silana
KHARKHAUDA
Halalpur - II
REFERENCES
State Boundary
District Boundary
Ancient Site
Modern Town

MAP-5

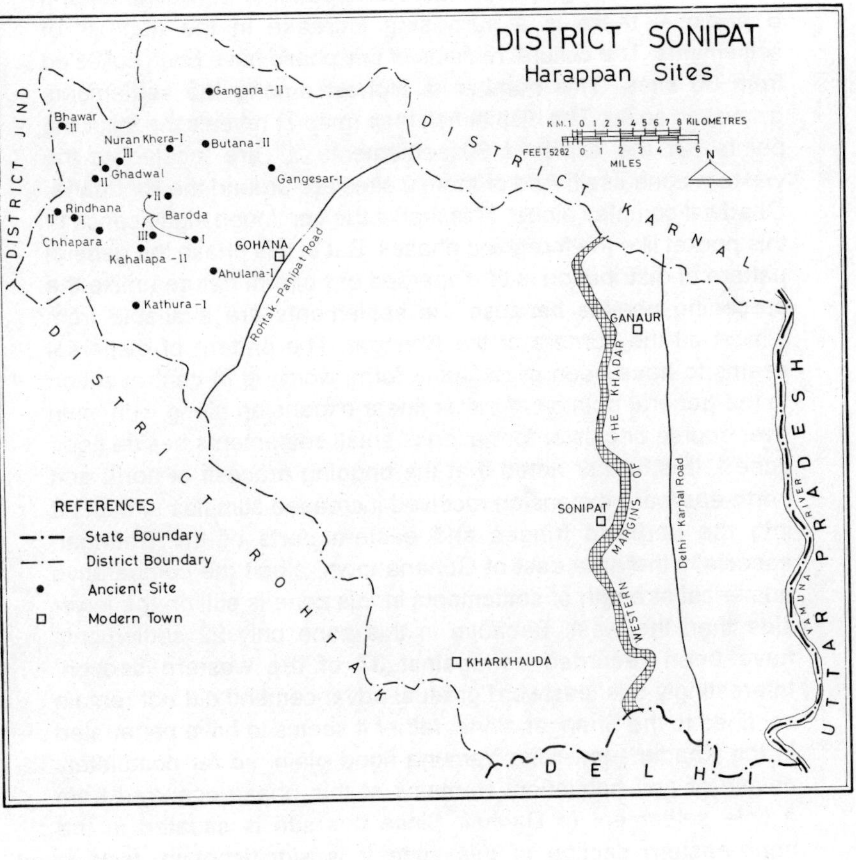
DISTRICT SONIPAT
Harappan Sites
K.M. 1 0 1 2 3 4 5 6 7 8 KILOMETRES
FEET 5282 0 1 2 3 4 5 6
MILES
N
DISTRICT JIND
DISTRICT KARNAL
DISTRICT ROHTAK
UTTAR PRADESH
DELHI
Gangana - II
Bhawar - II
Nuran Khera - I
Butana - II
Ghadwal
Gangesar - I
Rindhana
Baroda
Chhapara
GOHANA
Kahalapa - II
Ahulana - I
Kathura - I
GANAUR
SONIPAT
KHARKHAUDA
Rohtak - Panipat Road
Delhi - Karnal Road
THE KHADAR OF WESTERN MARGINS
YAMUNA RIVER
REFERENCES
State Boundary
District Boundary
Ancient Site
Modern Town

MAP-6

with the people as there is hardly any settlement. Likewise the eastern section also remained in oblivion so far as habitation and currency of cultural tradition of this phase is concerned.

In the following phase—late Harappan/late Siswal/Mitathal-II B phase — there is a surprising increase in the number of settlements. The cultural remains of this phase have been collected from 60 sites. This number is highest among the settlements discussed so far. The distribution map (map 7) reveals the following points. Of the explored 60 settlements 37 are located in the western zone itself. And of them 9 sites are around the Rindhana-Ghadwal complex alone. This marks the continued significance of this pocket like the foregoing phases. But in this phase the general pattern of distribution is of dispersed or uniform nature unlike the preceding phases because the settlements are available from almost all the corners of the Bhangar. The pattern of dispersal seems to have been of radiating form, which is in contravention to the general opinion of either linear expansion along with main river course or cluster formation of small settlements beside flood zone[24]. It is further noted that the ongoing process of north and north-eastward expansion received increased stimulus to expand into the northern fringes and eastern parts of the Bhangar, especially the area east of Gohana town, albeit the comparative numerical strength of settlements in this zone is still on the lower side than the west. Because in this zone only 22 settlements have been recorded as against 37 of the western section. Interestingly this 'eastward gradual advancement' did not remain confined to the Bhangar alone, rather it seems to have penetrated in the Khadar—low-lying Yamuna flood plain, so far completely devoid of any habitation. Remains of this phase occurred from a sole settlement of Datauli. Since this site is situated in the north-eastern section of this zone it is significant in itself in various ways. And this is not in isolation because occurrence of similar remains has been reported earlier by Amar Singh from the adjoining parts of district Karnal (now Panipat)—Haryana in the north[25]. This signifies the extension of habitations to this point in this phase. This growing tendency of breaking fresh grounds for making new settlements explains the pattern of distribution of settlements and migration of people, which further confirms the aforesaid observation of 'gradual eastward expansion' of human activities.

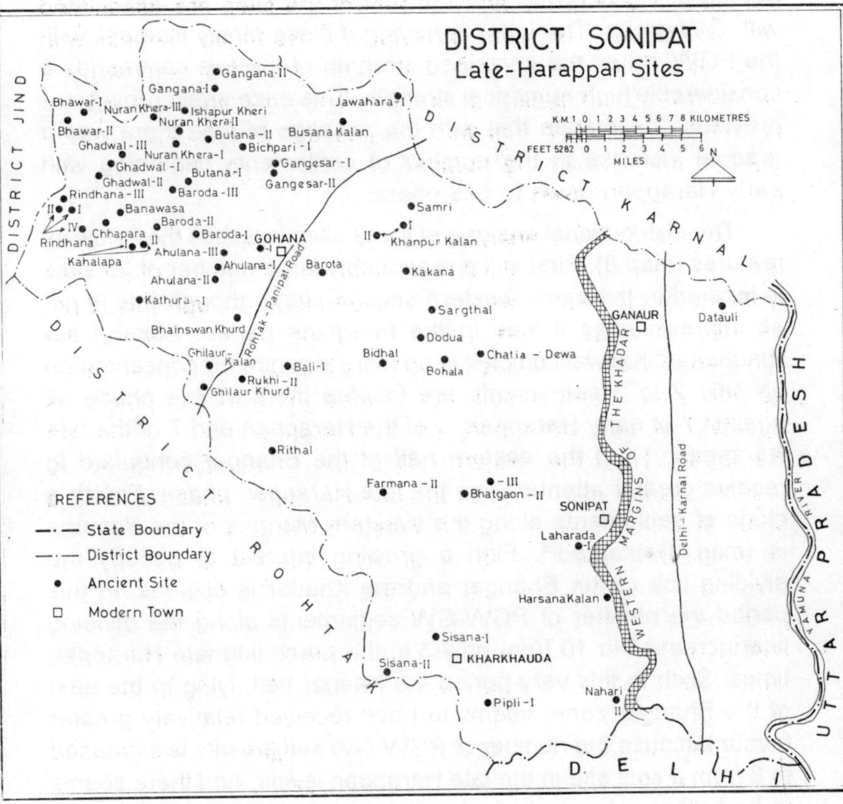
DISTRICT SONIPAT
Late-Harappan Sites
KM 1 0 1 2 3 4 5 6 7 8 KILOMETRES
FEET 5282 0 1 2 3 4 5 6
MILES
N
DISTRICT JIND
DISTRICT KARNAL
DISTRICT ROHTAK
DELHI
UTTAR PRADESH
YAMUNA RIVER
WESTERN MARGINS OF THE KHADAR
Delhi - Karnal Road
Rohtak - Panipat Road
Gangana-II
Gangana-I
Bhawar-I
Nuran Khera-III
Ishapur Kheri
Nuran Khera-II
Jawahara-I
Bhawar-II
Busana Kalan
Ghadwal-III
Butana-II
Nuran Khera-I
Bichpari-I
Ghadwal-I
Gangesar-I
Ghadwal-II
Butana-I
Gangesar-II
Rindhana-III
Baroda-III
Banawasa
Samri
Baroda-II
Chhapara
Baroda-I
Rindhana
GOHANA
Khanpur Kalan
Ahulana-III
Kahalapa
Ahulana-I
Barota
Ahulana-II
Kakana
Kathura
Sargthal
Bhainswan Khurd
GANAUR
Datauli
Dodua
Ghilaur-Kalan
Bidhal
Chatia-Dewa
Bali-I
Bohala
Rukhi-II
Ghilaur Khurd
Rithal
Farmana-II
Bhatgaon-II
SONIPAT
Laharada
Harsana Kalan
Sisana-I
Sisana-II
KHARKHAUDA
Pipli-I
Nahari
REFERENCES
State Boundary
District Boundary
Ancient Site
Modern Town

MAP-7

In the succeeding phase of PGW-GW we witness circulation of a novel pottery, which marks a different stage in cultural evolution. The cultural remains of this phase have been gleaned from 63 settlements of various dimensions. Of them 39 settlements belong to PGW group and the rest of the sites are associated with GW group. The latter is having a close family likeness with the PGW. Thus, the combined strength of the two commands a considerably high numerical strength. This once again proves our previous observation that with the passage of time there was a gradual increase in the number of settlements beginning with early Harappan down to this phase.

The distributional analysis of these sites indicates the following features (map 8). First still a reasonably large number of 23 sites is located in the same western section alone, though this is not as impressive as it was in the foregoing period. Second the Rindhana-Ghadwal complex is no more a centre for concentration as only 2 to 3 settlements are located there in this phase as against 7 of early Harappan, 4 of the Harappan and 7 of the late Harappan. Third the eastern half of the Bhangar continued to receive greater attention like the late Harappan phase. Fourth a chain of settlements along the Western Margins of the Yamuna is (map 4) marked[26]. Fifth a growing interest to occupy the dividing line of the Bhangar and the Khadar is obvious. In this period the number of PGW-GW settlements along the dividing line increases to 10 from only 3 in the preceding late Harappan times. Sixth in this very period the Khadar belt, lying to the east of the Bhangar zone, seems to have received relatively greater favour because the number of PGW-GW settlements is increased to 8 from a sole site in the late Harappan levels. And there seems to be further extension of a habitational process initiated only in the preceding phase. But interestingly, even in this phase the area of their frequency remained the northern section. This needs to be further elaborated in the next chapter on 'Settlement Patterns'. Next these sites are at a safe distance from the river Yamuna, i.e., none of them is near the active Yamuna watercourse. If we go by the sequence and spread of PGW-GW settlements, especially in this section of the Khadar, then the overwhelming presence of the sites is suggestive of posterior nature of the beginning of human habitation than that of the Bhangar in the west. It further denotes economic condition of the

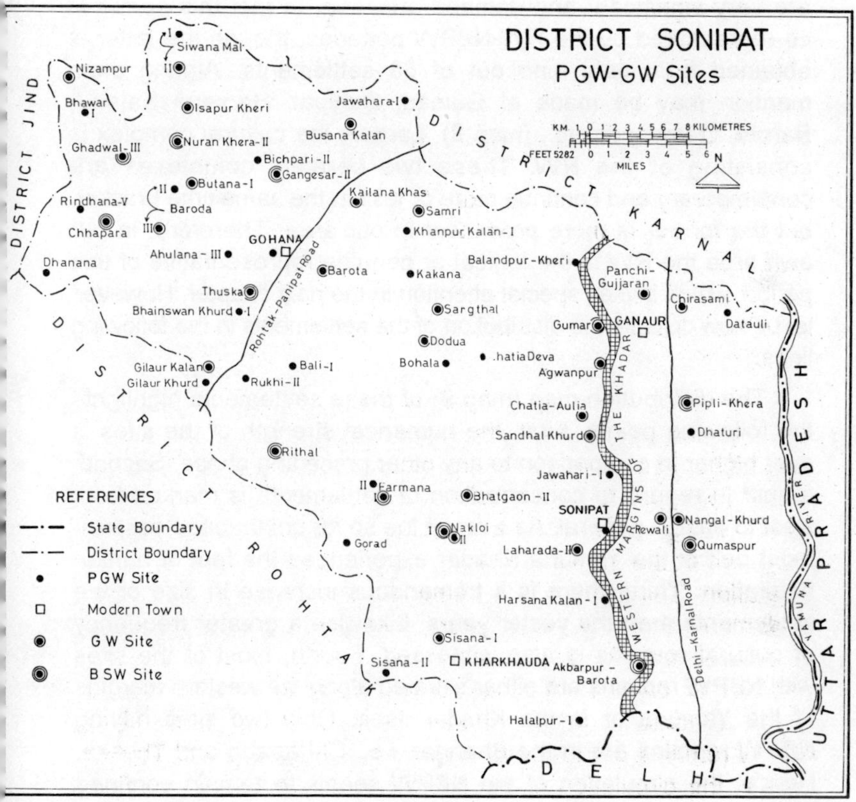

DISTRICT SONIPAT
PGW-GW Sites
KM 1 0 1 2 3 4 5 6 7 8 KILOMETRES
FEET 5282 0 1 2 3 4 5 6
MILES
N
DISTRICT JIND
DISTRICT KARNAL
DISTRICT ROHTAK
DELHI
UTTAR PRADESH
YAMUNA RIVER
WESTERN MARGINS OF THE KHADAR
Rohtak-Panipat Road
Delhi-Karnal Road
Siwana Mal
Nizampur
Bhawar I
Isapur Kheri
Jawahara-I
Nuran Khera-II
Busana Kalan
Ghadwal-III
Bichpari-II
Gangesar-II
Butana-I
Baroda
Rindhana-V
Kailana Khas
Samri
Chhapara
Khanpur Kalan-I
GOHANA
Ahulana-III
Balandpur-Kheri
Dhanana
Barota
Kakana
Panchi-Gujjaran
Thuska
Chirasami
Bhainswan Khurd-I
Sargthal
Datauli
Gumar
GANAUR
Dodua
Bohala
hatia Deva
Agwanpur
Gilaur Kalan
Bali-I
Gilaur Khurd
Rukhi-II
Pipli-Khera
Chatia-Aulia
Dhaturi
Sandhal Khurd
Rithal
Jawahari-I
Farmana
Bhatgaon-II
SONIPAT
Rewali
Nangal-Khurd
Nakloi
Gorad
Qumaspur
Laharada-II
Harsana Kalan-I
Sisana-I
Sisana-II
KHARKHAUDA
Akbarpur-Barota
Halalpur-I
REFERENCES
State Boundary
District Boundary
PGW Site
Modern Town
GW Site
BSW Site

MAP-8

people. Thus, human influence somehow, penetrated well inside the Yamuna Khadar covering sizeable northern parts.

Following the PGW-GW phase there appeared the RW-NBPW phase on the cultural horizon. The cultural remains of this phase are very significant and demand attention. In fact this period is co-represented by the RW-NBPW potteries, though the latter is obtained from only nine out of 88 settlements. Among them mention may be made of Gumad, Sonipat, Harsana Kalan-I, Barota, Qamaspur etc., (map 9). Largely the cultural complex is consisting of the RW. These two ceramic complexes are contemporary and continue more or less in the same time bracket, but the former is more prominent in our area. Therefore, in our own area the RW is the logical or genuine representative of this period, which invites special attention in the next chapter. However, let us now discuss the distribution of the settlements in the following lines.

The distribution map (map 9) of these settlements highlights the following points. First, the numerical strength of the sites is ever higher in comparison to any other preceding phase. Second, a shift in regard to concentration of settlements is marked from west to east in general. As a result the so far uninhabited eastern-most belt of the Yamuna Khadar experienced the feel of human habitation. Third, there is a tremendous increase in size of the settlements than the yester years. Likewise a greater frequency of cultural remains is also witnessed. Fourth, most of the sites with NBPW remains are either situated along the western margins of the Yamuna or in the Khadar itself. Only two sites having NBPW remains are in the Bhangar, i.e., Chhapada and Thuska. Hence, the circulation of the NBPW seems to remain confined mainly to the Khadar extension. Similar observation was made by Jagdish Kumar in connection with NBPW settlements in Haryana[27]. Lal also observed a higher number of NBPW sites in the Yamuna valley around Kanpur[28]. However, the foregoing presentation makes one point amply clear that is related with the process of gradual expansion of habitation towards east, which began sometime in the early Harappan levels and seems to have attained the highest mark in this phase by reaching to the very active flood plain of the Yamuna, of course, passing through various intervening stages in this process. This data on settlements can be compared with

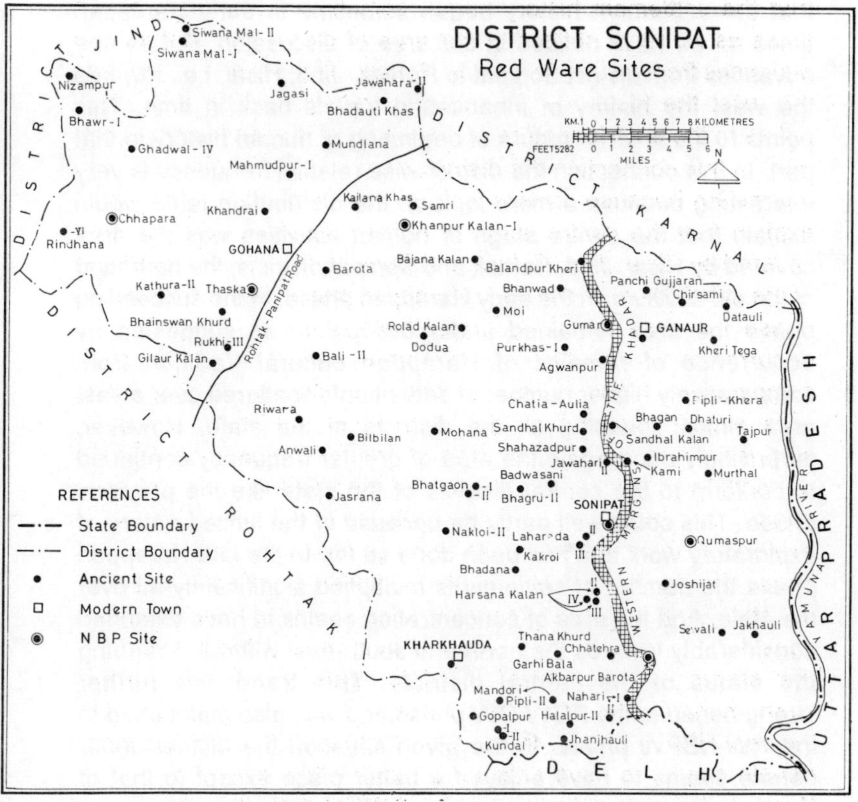
DISTRICT SONIPAT
Red Ware Sites
KM.1 0 1 2 3 4 5 6 7 8 KILOMETRES
FEET5282 0 1 2 3 4 5 6
MILES
N
REFERENCES
State Boundary
District Boundary
Ancient Site
Modern Town
N B P Site
DISTRICT JIND
DISTRICT KARNAL
DISTRICT ROHTAK
DELHI
UTTAR PRADESH
YAMUNA RIVER
KHADAR
THE WESTERN MARGINS OF THE KHADAR
Rohtak - Panipat Road
Siwana Mal - II
Siwana Mal - I
Nizampur
Bhawr - I
Jagasi
Jawahara - II
Bhadauti Khas
Mundlana
Ghadwal - IV
Mahmudpur - I
Kailana Khas
Samri
Khandrai
Chhapara
Rindhana
Khanpur Kalan - I
GOHANA
Bajana Kalan
Barota
Balandpur Kheri
Kathura - II
Thaska
Bhanwad
Panchi Gujjaran
Chirsami
Datauli
Moi
Bhainswan Khurd
Rolad Kalan
Gumar
GANAUR
Rukhi - III
Dodua
Purkhas - I
Kheri Tega
Gilaur Kalan
Bali - II
Agwanpur
Chatia - Aulia
Pipli - Khera
Riwara
Bhagan
Dhaturi
Sandhal Khurd
Mohana
Tajpur
Bilbilan
Sandhal Kalan
Anwali
Shahazadpur
Ibrahimpur
Jawahari
Kami
Murthal
Badwasani
Bhatgaon
Bhagru - II
Jasrana
SONIPAT
Nakloi - II
Laharada
Qumaspur
Kakroi
Bhadana
Joshijat
Harsana Kalan
Sevali
Jakhauli
Thana Khurd
KHARKHAUDA
Chhatehra
Garhi Bala
Akbarpur Barota
Mandori
Pipli - II
Nahari - I
Gopalpur
Halalpur - II
Jhanjhauli
Kundal

MAP-9

the reported number of settlements from other districts of the state in order to know the general frequency of sites in different cultural phases and the areas of concentration. This is given in Table 4.22. The distribution of settlements on district lines shows that the settlement history began sometime in early Harappan times as we have noticed in our area of discussion. But as one advances from district Sonipat to Rohtak, Jind, Hisar, i.e., towards the west the history of inhabitation travels back in time. This points to the anterior nature of beginning of human history in that part. In this connection the district-wise relative frequency is very interesting because a mere look on the distribution table would explain that the centre stage of human activities was the area covered by Hisar, Jind, Rohtak and Sonipat districts, the heartland of the old alluvium, in the early Harappan phase. In the succeeding phase the area remained under occupation as suggested by occurrence of remains of Harappan cultural tradition from comparatively higher number of settlements scattered over a vast area almost touching all the districts of the state. However, surprisingly once again the area of greater frequency continued to conform to the central districts of the state like the previous phase. This could well partly be because of the limited nature of exploratory work that has been done so far. In the late Harappan phase the number of settlements multiplied significantly all over the state. And the area of concentration seems to have extended considerably towards the north and south-east without disturbing the status of the central districts. This trend was further strengthened in the PGW-GW phase and was also maintained in the RW-NBPW phase. In the given situation the district under debate seems to have enjoyed a better piace except to that of Kurukshetra especially in respect of PGW-GW sites at present.

Locational Analysis

In the following pages it is proposed to make some observations regarding the process of site selection adopted in various periods by the people. Keeping in view the chronological sequence of cultural phases it would be appropriate if one begins with early Harappan settlements. In this connection it has been observed that all the sites belonging to early Harappan phase are situated on rising aeolian sandy pockets. Invariably these sites are also located beside a shallow natural depression with north-west to south-east orientation on an average, which usually corresponds to the general sloping tendency of the plains. Similarly most of the sites in Rajasthan[29], in Saurastra[30] and in the Bahawalpur region[31] are also found located in the same manner. In this regard there are some deviations also from the general regional pattern. To highlight this point one may mention some depressions of local nature with distinct orientations contradicting the regional patterns. The longer axis of these depressions are either in the east-west or north-south direction. This demonstrates the existing natural scene and its impact on the extension pattern of these settlements. However, these depressions received rain-water from the adjoining areas at local as well as regional scale and thus, served as main repositories for local and regional rain-water. The water thus, accumulated would have been retained for greater part of the year, if not the whole year, to cater water needs of the people through times. These very depressions seem to have served as the main lifeline in the absence of any other source of water, since the longer axis of the early Harappan settlements are from north-west to south-east oriented in consonance with the drainage pattern there is every likelihood of their being guided by the then prevailing conditions of availability of water. And that is why the orientation of settlements is in agreement with the drainage pattern of the area. In this phase the average size of settlements is very small and in all respect they appear like hamlets consisting only of a few families (a detailed discussion on this count is in the following chapter). The whole gamut of cultural deposition also does not sound a happy note because usually it is very thin, flimsy and tend to give a transient look. This has further been corroborated by highly restricted circulation of cultural artifacts, which are specimens of poor workmanship altogether as their surface is uneven and dull or largely being handmade. All these features tend to point to a simple life preoccupied mainly with subsistence needs.

Coming to the succeeding mature Harappan phase people exhibit a fairly strong liking for rising pockets irrespective of their being composed either of aeolian sand or of sandy loam soil. In fact what is interesting is the increasing preference for elevated aeolian sandy soils as witnessed in regard to Ghadwal–I; Rindhana–II, Baroda–I and III, Butana–I, Gangesar–I, etc., (map 6). As a rule these sites are always found located along or near some low-lying seasonal watercourse or depression of local as well as regional nature as was the situation in the previous phase. This is, of course, a feature of the preceding period in continuation. Nevertheless there are some sites like Bhawad–II and Kathura-I, etc., found located on almost level soil pockets, though in such cases there is a strong possibility of their being elevated previously because these are now under intensive cultivation. Such a tendency among the people of selecting rising pockets would have been caused by local topographic conditions, irregular behaviour of rainfall and drainage lines as well and more so in absence of matching operative potential. Most of these settlements are found enlarged in a slip like fashion in accordance with the existing orientation of the natural depressions, whether be it with north-west to south-east or east-west or north-south orientation. However, this could well be a case of what rain and surface flow have caused to the mounds over the years. Interestingly most of the settlements are found extended in the north-west to south-east direction in accordance with the general sloping pattern of the area, which is a continuation of the early Harappan practice here. Besides there are also examples of contradicting the general patterns. In this context one may quote the east-west orientation of the site of Baroda–III. It is because the natural depression to the north of the site is stretched in the similar directions. Such trends are quite obvious even these days in respect of modern villages. Here village ponds tend to assume similar role. This reveals on the one hand contrasting, but interacting geographical and ecological features at microlevel and on the other hand their overwhelming predominence over man who were handicapped in respect of technological know-how and organizational strength. This provides insights into the *modus operandi* of human society to survive through adverse natural conditions.

In connection with the size of the settlements we witness surely a general tendency of enlargement at almost all the sites of this phase. So is the case in relation with thickness of cultural deposits at some of the better conditioned settlements and on an average as well. This, however, points to the greater time span of development enjoyed by these levels. Moreover the process of advancement is further elaborated by greater frequency of artifacts, especially pottery of better quality at almost all the settlements of this phase. These aspects sound an element of greater sedentariness and stability in their life. Therefore, we can simply suggest here an improvement over the preceding phase. The details of, which are to be dealt with in succeeding chapter.

In the late Harappan/Mitathal–II B levels the theme of site selection once again projects a continuation of the so far ongoing trend, since large number of settlements is recorded on aeolian rising sandy pockets. Some of these settlements are also noticed on now level grounds, such as Rukhi-II, Bohla, Busana and Jawahara-I. Since these sites are now under intensive cultivation, there is every possibility of their being placed on elevated pockets before levelling for cultivation. And this fact has further been borneout by versions of the villagers. To justify this point further one could very well take note of the reportings by Silak Ram. In this connection he mentioned a reasonably big size of the site of Rindhana–I in 1972[32], whereas, when I visited the site in 1985, only a few remains were found just to mark the existence of a site. Such a situation is witnessed at several other sites also. In some cases it is also observed that the villagers have dug up sand from ancient mounds reducing them to surface level. However, the fact remains that the overwhelming number of settlements is located on rising sandy pockets.

With regard to their system of expansion a continuity of the earlier trend of having north-west to south-east orientation is reaffirmed. In this phase also most of the settlements are seen extending along the direction of water. But these are away from the river courses like the preceding ones. Similar observation was made by Bhan[33]. And that is why there are some other sites with different orientation, such as north-south and east-west, which is primarily again based on the flow of the water in the adjacent natural depression. This highlights two important points. First that the choice of site selection was still under constant influence of local geographical conditions in an ongoing environment of technological constraint. Second is that there existed considerable variations in the ecological and topographical realms at microlevel as against general presumption of homogeneous and striking uniformity over the area in these respects. This type of conducive combination of natural forces would have unfolded a wide range of subsistence activities to the protohistoric people.

In regard to the size of settlements, their cultural thickness and the frequency of artifacts one can safely conclude in the following manner. The size of the settlements on an average is very small comprising only a few hundred square metre. And so far as the cultural thickness is concerned the sites are usually of low-lying nature with very thin or at times with no deposition at all reducing them to the ground level. This bears similarity with the early Harappan scenario in this context. And as regards the frequency of cultural artifacts it is highly limited except pottery, which is available in abundance to predominate the cultural composition, but displays a devolutionary process. However, there is hardly any sign of cultural discontinuation. In this regard the characteristic features of the mature Harappan pottery are conspicuous by their absence despite the fact that the pottery is of same group. Interestingly, a recurrence of some of the early Harappan traits, especially in case of ceramic tradition, in the form of an underlying current is also experienced. It seems that the influence of mature Harappan period was not strong enough to replace the early Harappan traditions altogether in this eastern rural belt. And the moment mature Harappan urban influence began to fade off some of the surviving rural traits of the early phase became quite obvious.

Coming to the PGW-GW phase we encounter an interesting situation. In this period the settlements are found located in almost all types of locales such as level and raised grounds. However, usually aeolian sandy pockets were preferred to over level and low-lying pockets in the Bhangar. This process of site selection bears a close affinity with that of the preceding phases. Even in Yamuna Khadar specially the settlements are located on rising alluvial fans well above the reach of the Yamuna floods. All these settlements, both in the Bhangar and the Khadar, are located close to some source of water in the form of a natural depression. Sizewise most of the settlements are small like the late Harappan sites. But interestingly, some of them demonstrate a strong tendency of enlargement as at Bhawad-I, Gangesar-II, Khanpur-I, Rithal in the Bhangar (around Gohana town) and Gumad, Agwanpur, Sandhal Khurd, Jawahari, Harsana Kalan-I, Akbarpur Barota, etc., along western margins of the Yamuna Khadar. And some of the settlements in the Khadar also seem to have assumed that position. This trend of aggregation is significant and envelopes several mute points, which shall be taken up for detailed discussion in the succeeding chapter on 'settlement patterns'. Hence, it seems after the breakup of the Harappan system once again in this phase the size of sites underwent an evolutionary transformation. In the same manner similar trends have also been recorded with regard to their depositional thickness and frequency of cultural artifacts at the aforesaid settlements, while some others maintained their low-elevation. And still many others tended to be almost flat with no depositional elevation whatsoever. The cultural repertoire does not seem to be very rich if we exclude pottery, especially the PGW.

In succession to PGW we witness emergence of a novel tradition known as RW-NBPW. The cultural remains of this period are significant in the sense of their cultural composition, wherein admixture of the RW and NBPW is available, though the latter remains are obtained only from 9 settlements as mentioned before. The scanty remains of this pottery are indicative of their symbolic presence in the overall cultural milieu. Largely the cultural complex is consisting of RW. This ware, therefore, being contemporary of NBPW is the real representative in this area.

The map 9 with the sites of this phase highlights certain points as given below. First, of all the number of settlements is very large and fairly spread over the entire zone. Second, a growing shift in regard to concentration of settlements is marked from the nuclear west zone to the east zone. Third, the so far uninhabited easternmost belt of the Khadar, along the active river course in the west, is occupied for the first time only in this period. Fourth there is a tremendous increase in size of the settlements over the past periods. And similarly in line with this general trend a greater frequency of cultural artifacts is also appreciable. Thus, in this phase an all-round improvement is quite obvious, which calls for further elaboration in the coming chapter.

Discussion

The foregoing discussion on the process of site selection, size of settlements and their cultural composition points out frequently an evergrowing tendency of the people to adopt rising sandy pockets either composed of aeolian or alluvial sand through times, though occasionally elevated loam and clayey pockets were also selected. This must have been based on experience and experiments of generations in a given area. This invokes some apprehensions as usually to why people tended to opt for such pockets? To explain it we may elaborate in the following manner that these people preferred to such pockets for their being safe from floods, wild animals and contemporary human beings. Besides on such pockets there was a distinct possibility of effects of approaching salinity as such soils were already being generally deficient in salt contents. These soils are highly porous in nature being composed of predominant sand particles and as a result enjoy high water absorbing capacity. Therefore, such soils soak rain as well as other water quickly and thus, are capable of providing instant dry ground for human beings and animals, even in rainy season than the level, low-lying and soils constituted with high content of salt. This could have proved very hygenic both for human beings and animals. Moreover such soils hardly suffer from rising water-table and consequent waterlogging conditions. The pH of such soils usually ranges between 7 to 8.5, which is the optimum range[34]. And perhaps keeping in view such properties of these soils people usually term them as 'thandi dharati'—cool soils. Such soils do enjoy tremendous amount of climatic shock absorbing potential. The shock of this nature are not very uncommon given the quick and abrupt fluctuations in the realm of atmospheric temperature of this area. The capacity to absorb such ups and downs helps check adverse effects on the maturing process of crops and their yields. Moreover these soils being loose in nature are known as self-ploughing and fertilizing.[35] This helps not only to grow even dry crops—'barani' without any extra water for irrigation, but also with the help of then available simplest tools and techniques. Moreover such soils demand much less tilling and physical labour to prepare seed-bed for sowing crops. There was perhaps hardly any need for large-scale clearing as such soils support only sparse and short vegetation of dry nature, which is so distinctly shown by the near absence of humus content on the surface soil. Such pockets would have also provided with ample opportunities for open and direct sun during winter and unhindered fresh air during summar nights. Similarly there was better possibility of aeration in these soils, which goes a long way not only to maintain, but also to increase soil fertility, so crucial a factor for maintaining pace of agricultural production. Moreover penetration of sun rays into the crop root zone is equally essential for better growth and yield. And this is usually feasible in open and soft soils. Thus, these soils were most suitable for them and perhaps that is why people went for them.

Conclusion

The foregoing discussion brings out certain points as given below. The beginning of food production in this part of the belt coincided with appearance of early Harappan or early Siswal-B settlements sometime in the second half of the early Harappan phase. Once this process was initiated with limited number of small sized settlements of early Harappan phase, the sites gradually registered a growth in terms of size, number, depositional thickness and cultural advancement right from early Harappan down to RW-NBPW levels with an interlude in late Harappan levels. But that too centred around the size of sites and cultural richness only. Second the process of inhabitation experienced an eastward extension by way of gradual eastward advancement right from early Harappan down to RW -NBPW levels. In the last phase this attains the highest stage so far as the district is concerned by way of making settlements well along the easternmost point of the district, the western bank of the Yamuna river. Whereas in the first two stages of inhabitation the process did not experience any tangible change as their area of interaction basically remained confined to the western pocket of the district, i.e., west of Gohana town. The second stage of this process is related with the late Harappan phase in which a striking extension of the ambit of their area of influence spreads over the entire Bhangar zone. This highlights fresh occupation of those eastern areas of the Bhangar, which hitherto remained beyond parameters of human habitation. And this eastward expansion was not confined to the Bhangar belt alone any longer as this transcended its limits by making inroads into the alien Khadar zone for the first time in this phase if late Harappan remains from Datauli are any indication. Since this site is located in the north-eastern pocket of this district, this suggests north-east orientation of their 'gradual process of eastward advancement'. In the third stage the continuation of the ongoing eastward movement is witnessed to envelope almost all the remaining parts of the northern half of the Khadar. In this very period of PGW-GW a vertical shift in their area of frequency from west to east is recorded. And the area alongside the Western Margins and the northern half of the Khadar itself becomes a centre of increasing activities. And finally in the last stage consisted of RW-NBPW the whole of the Khadar except the present water-course of the Yamuna is touched upon in one way or the other for habitation purposes. In this very period a greater frequency of settlements is witnessed in the eastern section. This in itself is indicative of an ongoing process of shifting emphasis from west to the east, which broadly forms a part of the process of 'gradual eastward expansion'. Thus, the process triggered off sometime in the early Harappan levels seems to have attained its watermark in the early historical period with the emergence of a relatively large number of settlements in the eastern half, the flood plain of the Yamuna. As to what caused this development is to be discussed in Chapter IV.

NOTES AND REFERENCES

1. Dutt, B., *Painted Grey Ware Settlements in Haryana*, Ph.D. Thesis (unpublished) Kurukshetra University, Kurukshetra, 1980.
2. Phogat, S.R., *Archaeology of Rohtak and Hisar Districts of Haryana*, Ph.D. Thesis (unpublished), Kurukshetra University, Kurukshetra, 1972.
3. Thakran, R.C., 'Change and Continuity of a Protohistoric Tradition in the Sutlej-Yamuna Divide', *PIHC*, Srinagar, 1986, pp. 150-60. Here Siswal B of Suraj Bhan is termed early Siswal B.
4. Bhan, S., *Mitathal Excavations and Other Explorations in the Sutlej-Yamuna Divide*, Kurukshetra University, Kurukshetra, 1975.
5. *Ibid.*
6. *Ibid.*
7. Thakran, 1986.
8. The foregoing culture-wise breakdown of explored sites displays seemingly an inflated number than the factual strength recorded. This is mainly because several settlements represent either double or multiple cultural complexes. With regard to this work these multicultural remains at one site are taken as independent representative of respective cultures.

9. Bhan, 1975.
10. Bhan, S., 'Archaeological Evidence of the Changes in the Course of the Yamuna in Sub-recent Times', *JHS*, II, 1-2, Kurukshetra University, Kurukshetra, 1970, pp. 1-3.
11. Bisht, R.S., 'Banawali: a New Harappan Site in Haryana', *ME*, II, Delhi, 1978. pp. 87-88; Idem. 'Excavation at Banawali: 1974-77', in Possehl, G.L. (ed.), *Harappa Civilization* I, New Delhi, 1982, p. 161; Idem. 'Structural Remains and Townplanning of Banawali', in Lal, B.B. and S.P. Gupta (eds.), *Frontiers of Indus Civilization*, Delhi 1984, pp. 89-98, Idem. 'Banawali and Some Other Recent Excavated Harappan Sites in India', in Taddi, M. (ed.), *South Asian Archaeology*, 1977, Naples 1979, pp. 223-40.
12. Joshi, J.P., 'Life During the Overlap of Late Harappa and PGW Cultures', *Journal of Indian Society of Oriental Art*, IX, 1978, p. 22; Idem. 'New Light on the Post-Urban Phase of the Indus Civilization in India', paper presented at the Seminar *On the Harappa Civilization in India*, held at Srinagar from June 22-26, 1979; Idem. 'A Note on the Excavation at Bhagwanpura', *Puratattva*, New Delhi, 1976, p. 179; Idem. 'A Note on the Excavation at Bhagwanpura', in Gupta, S.P. (ed.), *Mahabharata : Myth and Reality*, New Delhi, 1976, pp. 238-39; *IAR*, 1975-76, p. 16, Idem, *Bhagwanpura Excavations,* ASI, New Delhi, 1993.
13. *IAR*, 1965-66, pp. 35-36.
14. *IAR,* 1970-71, p. 15.
15. Dutt, 1982, p. 192.
16. *IAR*, 1970-71, pp. 15-16.
17. Bhan, S., 'Srughan or Sugh—An Old Capital of Ancient Punjab', *VIJ*, V, Hoshiarpur, 1967, pp. 84-89.
18. Dutt, 1982.
19. Ammerman, A.J. and L.L. Cavalli-Sforza, 'Population Model for the Diffusion of Early Farming in Europe', in Renfrew, Colin (ed.), *The Explanation of Culture Change: Model in Prehistory*, Duckworth 1972, pp. 343-57.
20. Agrawal on the basis of C-14 dates assigned a time bracket between c. 2400-2000 BC to Kalibangan-I (Agrawal 1971: 116-18). Suraj Bhan traced a cultural similarity between the upper level of Kalibanga I and the early levels of Mitathal-I. With this comparability he associated the beginning of Mitathal-I with the end of Kalibangan-I. As a result the beginning of Mitathal I coinciding with the end of Kalibangan-I was placed around c. 2000 BC which continued, according to him, for a century for the depositional

thickness of this phase is not more than a metre. Its end was, thus, placed at c. 1900 BC and the Mitathal-I phase as such in between c. 2000-1900 BC (Bhan 1975: 15). If we, however, take into consideration the end period of Kalibangan-I c. 2000 BC and the early date c. 2150 BC of Kalibangan-II as proposed by Agrawal there is an obvious overlap of 150 years between the upper levels of Kalibangan-I and the inception of Kalibangan-II. This very implied period of overlap forms, in fact, a transitional stage between the two events which is so obvious in Mitathal-I levels. We have already mentioned that Suraj Bhan noticed a marked similarity between the cultural composition of the upper levels of Kalibangan-I and to those of the Mitathal-I/Siswal B (which is being termed as early Siswal B here). Mitathal-I and early Siswal B, thus, being contemporary to the implied transitional phase at Kalibangan are placed between c. 2100-2000 BC And accordingly we propose to assign a period from c. 2000-1800 BC to Mitathal-II A complex.

21. Phogat, 1972, p. 55.
22. *Ibid.*, p. 46.
23. Lata, S., *Archaeology and History of Sampla Block, District Rohtak (Haryana)*, M.Phil. Dissertation (unpublished), Maharshi Dayanand University, Rohtak, 1987-88, pp. 26-27.
24. Bhan, S., 'Changes in the Courses of the Yamuna and their Bearing on the Protohistoric Cultures of Haryana', in Deo, S.B. (ed.), *Indian Archaeological Congress and Seminar papers, Nagpur University* 1972, pp. 125-28; Idem, 'Archaeological Evidence of the Changes in the Course of the Yamuna in Sub-recent Times', *JHS*, 11, 1-2, 1970, pp. 1-3.
25. Singh, A., *Archaeology of Karnal and Jind Districts of Haryana*. Ph.D. Thesis (unpublished), Kurukshetra University, Kurukshetra 1981, pp. 66-67.
26. Thakran, R.C., 'Western Margins of the Yamuna Khadar and the Protohistoric Cultures', *PIHC*, Dharwad, 1988, pp. 649-57
27. Kumar, J., *NBPW Settlements in Haryana*, M.Phil., Dissertation (unpublished), Kurukshetra University, Kurukshetra, 1980, pp. 80-81.
28. Lal, *op. cit.*, p. 55.
29. Ghosh, A., 'The Rajasthan Desert—Its Archaeological Aspects', *BNISI*, I, 1952, p. 42.
30. Chitalwala, Y.M., 'Harappan and Post-Harappan Settlements in Saurastra', in Agrawal, D.P. and B.M. Pande (eds.), *Ecology and Archaeology*, 1976, p. 93; Idem, 'Harappan and Post-Harappan

Settlements in Saurastra', in Agrawal, D.P. and D.K. Chakrabarti (eds.), *Essays in Indian Protohistory*, 1979, p. 114.

31. Mughal, M.R., 'Archaeological Evidence from Bahawalpur', *ME*, IV, Delhi, pp. 93-94.

32. Phogat, 1972, pp. 37-38.

33. Bhan, 1972.

34. Personal communication with S.S. Phogat, Incharge of Soil Chemistry Laboratory at Sonipat.

35. Hodges, H., *Technology in the Ancient World*, New York 1970, pp. 51-56.

4

SETTLEMENT PATTERNS

It has been marked in the preceding Chapter III that the area under consideration, forms part of the vast fertile plains of northern India, attracted the early farming communities way back sometime in the later half of the third millennium BC. Subsequently the area remained under occupation, albeit with interspersed gaps, till date to present several cultural strands and their stages. However, the present chapter deals with various other aspects of settlement systems in the said period. And the analysis is based on the available published as well as newly explored data. The inferences drawn in connection with dispersal of communities over the landscape should be treated tentative and subject to subsequent verifications through excavations. An attempt to analyze the locational process of these settlements is as follows.

Locational Analysis

This approach to settlement studies is one of the essential aspects help enhance understanding of the problem. As a result of growing interest and information, though still inadequate, the significance of such studies is increased and especially when the data regarding palaeogeographical, palaeoecological, palaeohydrological, palaeogeomorphological processes and cropping systems is awfully scant the utility of such attempts is further magnified. In such conditions the prevailing processes in regard to aforesaid aspects can serve as a guide to the ancient past, but surely in a limited manner[1]. Therefore, let us now view the available data in the light of present administrative units, geographical conditions, rainfall trends, hydrographic regime, cropping pattern, etc., to understand past human interactions against natural forces.

Tahsil-wise Distribution

As the district is consisting of two tahsils—Sonipat and Gohana, it is reasonable to classify the data on tahsil basis and explore the probable lines of dispersal of settlements. This has been given in table 4.1.

TABLE 4.1

Tahsil-wise Frequency of Protohistoric and RW-NBPW Sites

Culture	*Gohana*	*Sonipat*	*Total*
Early Harappan	10 (83.33%)	02 (16.67%)	12
Harappan	17 (100%)	00 (00%)	17
Late Harappan	46 (76.66%)	14 (23.34%)	60
PGW-GW	31 (49.20%)	32 (50.80%)	63
RW-NBPW	29 (32.09%)	59 (67.91%)	88
Modern Villages	94 (26.95%)	254 (73.05%)	348

The above table demonstrates the following points. In early Harappan phase both the Sonipat and Gohana tahsils support 2 and 10 sites respectively. By the time the mature Harappan phase appears an increase, though marginal, is recorded in their numerical strength rising to only 17. And all the settlements of this phase are located in tahsil Gohana. A close view of the process of site selection of the early and mature Harappan reveals a common approach as settlements in both the phases are found overlapping in the same western section with 9 and 17 sites respectively, i.e., the area west of Gohana town. It is important to note that the mature Harappan site distribution is relatively dispersed and more even than the early Harappan as indicated by maps 5 and 6. Similar tendency of overlapping has been attested to in adjoining areas of district Ambala and Kurukshetra[2], Karnal and Jind[3], Rohtak and Hissar[4].

Coming to the late Harappan levels we witness a marked increase in the numerical strength as indicated by a large body of 60 settlements. Interestingly 46 of them, however, belong to

Gohana tahsil and only 14 represent tahsil Sonipat. This remarkable increase in number of settlements surpasses both the previous phases. However, previously observed trend of site concentration in Gohana tahsil is reinforced. Like the earlier periods the centre of gravity seems to have been once again the same western section of Gohana. In the eastern section of Gohana and the contiguous western section of Sonipat tahsil a sporadic distribution of sites is noteworthy (map 7). Similar increase in number of sites is recorded in other parts of Haryana, Punjab and western Uttar Pradesh in this phase[5].

Subsequently PGW-GW remains seem to have followed the late Harappa culture in this zone (map 8). The remains of this culture have been gathered from 63 sites and their tahsil-wise distribution shows 31 to Gohana and 32 to Sonipat. It is true that the numerical strength of settlements in this phase is still higher than the number of preceding phases in general and the late Harappan in particular. However, the process of dissemination of habitations adopts a balanced position because in both the tahsils the respective number of settlements is almost even. But surely for the first time a shift in concentration from Gohana to Sonipat tahsil is visible as indicated by the respective numerical strength of the two tahsils (map 8). A process began in late Harappan phase takes a considerable leap in PGW-GW phase. This is further stimulated in RW-NBPW as out of 88 sites 29 and 59 are shared by Gohana and Sonipat tahsils respectively. And if we take it further to modern times two tahsils harbour 94 and 254 modern villages. This signifies a gradual eastward advancement of early farming communities on the one hand, which registers a high watermark in RW-NBPW levels, and on the other highlights the growing preference for tahsil Sonipat in the east with the passage of time, which receives highest number of inhabitants in recent times also (fig. 4.1). This is now, especially the easternmost belt, most thickly inhabited zone compared to other pockets in the west. This would help understand and explain the changing levels of densities in the two tahsils in ancient past.

Frequency of Sites in Physiographic Zones

The Sonipat district consists of two main physiographic zones

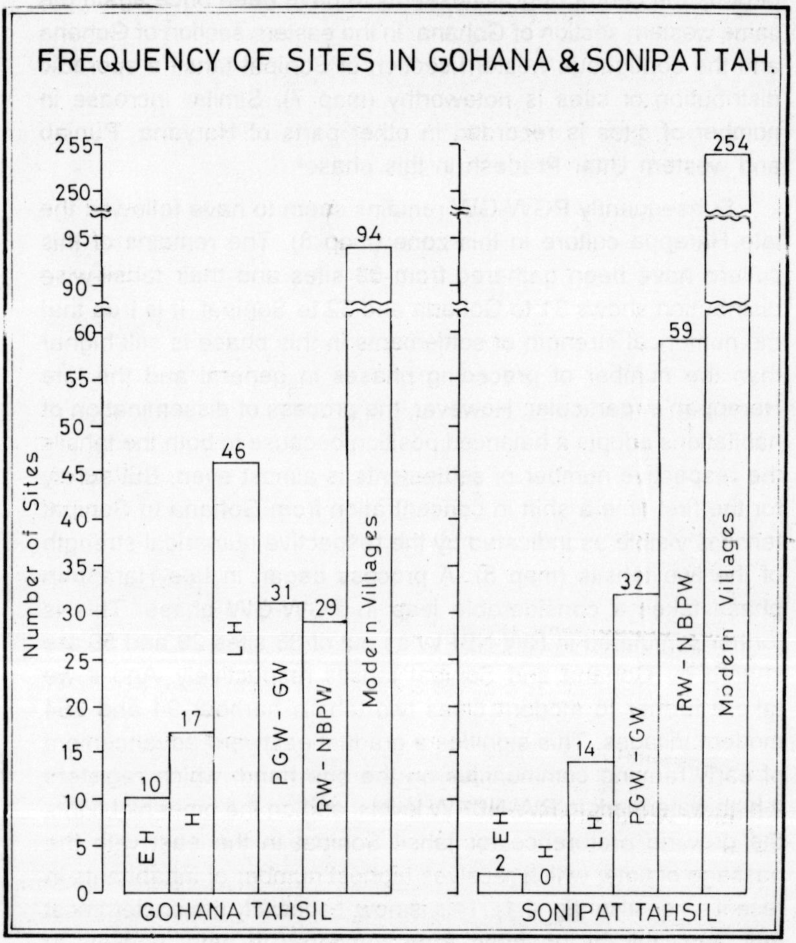

FREQUENCY OF SITES IN GOHANA & SONIPAT TAH.
Number of Sites
0
5
10
15
20
25
30
35
40
45
50
55
60
90
95
250
255
10
EH
17
H
46
LH
31
PGW - GW
29
RW - NBPW
94
Modern Villages
GOHANA TAHSIL
2
EH
0
H
14
LH
32
PGW - GW
59
RW - NBPW
254
Modern Villages
SONIPAT TAHSIL

Fig. 4.1

namely the Bhangar and the Khadar as has already been mentioned in Chapter I. These two are located in the west and the east respectively stretching side by side in north-south direction. The former covers the whole Gohana and western parts of Sonipat tahsil, and the Khadar embraces only eastern part of tahsil Sonipat. The data are analyzed according to these zones in the following lines.

The table 4.2 reveals that in the first two phases of the Harappa civilization the Khadar remained untouched because all the 12 and 17 settlements of early and mature Harappan phases were located in the upland dry Bhangar belt (fig. 4.2). Interestingly in late Harappan times the hitherto uninhabited low-lying Khadar was also penetrated as demonstrated by such remains from a sole settlement of Datauli in the north-east. The remaining 59 sites are dispersed over the Bhangar region alone. Similar remains of late Harappan phase have also been reported by Amar Singh from the adjoining southern parts of Panipat district.[6] Thus, the process of inhabitation of the Khadar seems to have started in late Harappan times witnessed greater number of settlements in the succeeding PGW-GW levels, which rises to 8. Obviously we experience a higher number of habitations in the Khadar, which explains the growing interest of inhabitants in this zone (fig. 4.2). But, in fact, the Bhangar still maintains its superiority over the Khadar. Similar observation was also made by Dutt.[7] This unfolds new vistas for deliberations on emerging settlement patterns and their stimulants. Coming to the RW-NBPW levels the number of sites increases further as the respective zones support 69 and 19 settlements. And even today so far as the number of villages is concerned the Bhangar is on top with 237 villages, whereas the Khadar contains only 111 villages. But, when the number of settlements is compared in respect of the size of the two zones, the Khadar being smaller in area surely enjoys superiority so far as the overall density of the modern settlements is concerned. How did it come is a very thought provoking point and could prove of immense help to make out the lost processes of sequence and spread of settlement sites here.

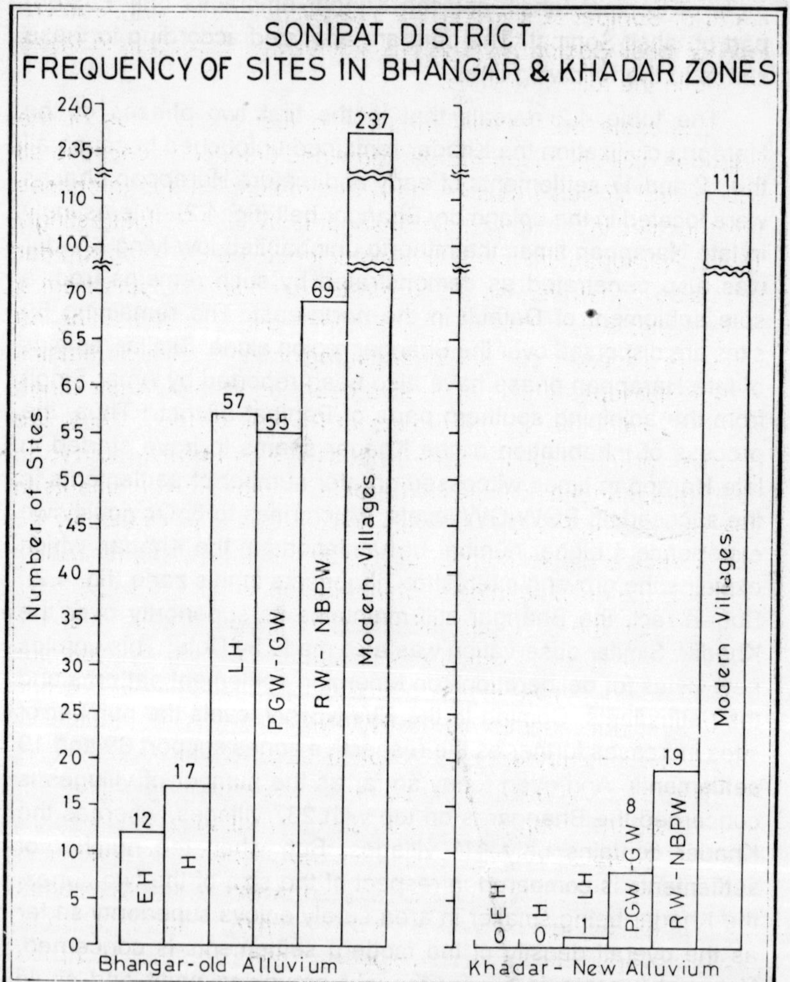
SONIPAT DISTRICT
FREQUENCY OF SITES IN BHANGAR & KHADAR ZONES
Number of Sites
240
235
110
100
70
65
60
55
50
45
40
35
30
25
20
15
10
5
0
12
EH
17
H
57
LH
55
PGW - GW
69
RW - NBPW
237
Modern Villages
Bhangar - old Alluvium
0
EH
0
H
1
LH
8
PGW - GW
19
RW - NBPW
111
Modern Villages
Khadar - New Alluvium

Fig. 4.2

Frequency of Sites in Relief/Physiographic Sub-Zones

TABLE 4.2

Frequency of Protohistoric and RW-NBPW Sites in Relief/Physiographic Zones

Sub-zone	*EH*	*H*	*LH*	*PGW-GW*	*RW-NBPW*	*Modern Village*
Bhangar West	09 (75.00)	14 (82.35)	37 (61.66)	23 (36.50)	20 (22.72)	52 (14.9)
Bhangar East	03 (25.00)	03 (17.65)	19 (31.66)	22 (34.92)	31 (35.22)	150 (42.75)
Bhangar	12 (100.00)	17 (100.00)	56 (93.32)	45 (71.42)	51 (57.94)	202 (57.65)
W. Margin	(00.00)	(00.00)	03 (05.00)	10 (15.87)	18 (20.48)	35 (10.05)
Khadar West	(00.00)	(00.00)	01 (1.68)	08 (12.71)	15 (17.04)	67 (19.2)
Khadar East	(00.00)	(00.00)	(00.00)	(00.00)	04 (4.54)	44 (12.6)
Khadar	(00.00)	(00.00)	01 (1.68)	08 (12.71)	19 (21.58)	111 (31.8)
Total	12	17	60	63	88	348

Note : Bracket figures are read as percentages.

For the purpose of this analysis the aforesaid two major geographical zones (Bhangar and Khadar) would further be divided into western and eastern subzones. This would perhaps help further understand the distribution pattern of ancient settlements on micro-level by providing us with minute details and insights into the factors governing the choice of site selection as shown in table 4.2.

The aforesaid distribution shows 9 of 12 early Harappan settlements in the western Bhangar and only 3 in the eastern half. This suggests significance of the western zone for the early farming communities who broke fresh grounds here and subsequently dispersed eastward in a scattered manner as indicated by 3 sites there. This is also indicative of the fact that the direction of gradual dissemination of habitation is from west to east. And the first occupation of this western pocket by the early Harappan people is a part of that process. With the beginning of habitation in the early Harappan levels the western subsection remains to have been a focal point in the entire Harappan levels, if 17 mature Harappan and 37 late Harappan settlements are any pointer. In PGW-GW levels the ongoing predominance of the western pocket is eroded considerably because the share of sites of both the pockets is (23 and 22 respectively), if 10 sites of the western margins are excluded from the eastern zone as shown by the distribution (table 4.2). This is indicative of a gradual shift with the beginning of the process of dispersal of human habitations from west to east. And what we witness in the succeeding phase of RW-NBPW is a further confirmation of that trend. In this period 31 sites recorded in the eastern subsection as against 20 in the western subsection of the Bhangar, if the 18 sites of the western margins are included in the eastern subzone their cumulative strength assumes very significant proportion of 55.70 per cent of the total strength. And even, if the share of the western margins is viewed against that of the eastern subzone it is equally important to take note of. This point is further elaborated if we take into account the respective number of 52 and 185 modern villages of the two subzones. Thus, the process of shifting focal point began sometimes in late Harappan times and seems to have accomplished in RW-NBPW phase.

In the Khadar, the second major physiographic zone, we confront a peculiar situation of beginning of human history with the occupation of eastern subzone of the Khadar sometime in the late Harappn phase as the sole late Harappan settlement of Datauli falls within this zone. The presence of this settlement in the north-eastern section of the district and the Khadar as well does not seem to have directly connected with the observed ongoing process of a gradual dispersal from west to east for the site is not contiguous with other explored late Harappan sites in the eastern Bhangar. Since there is a territorial gap between the sites in the eastern Bhangar and the site of Datauli in the north-eastern section of the Khadar, it becomes difficult to explain, correlate and locate in the given situation of the area under review. And if we go beyond the limits of our area of discussion it would be perhaps reasonable to explain and understand its location. To the north of our area lies the district of Panipat where a late Harappan site of Dikadala is reported.[8] This site is hardly 5-6 kilometre north-west of the site of Datauli in our area. Therefore, a correlation between the two sites is easy to establish. And the extension of late Harappan influence in this part of our district seems to have been a part of that process. This once again reaffirms our observation of gradual expansion towards east. However, the western section experiences the feel of human activities only in the succeeding PGW-GW phase as 8 settlements of this phase are located here only. In the RW-NBPW phase the number of settlements increases to 15, whereas the eastern section managed to attract people at 4 sites alone in RW-NBPW phase. This distribution brings forth the point of gradual expansion of farming communities from the west to the east, which began taking shape sometimes in the late Harappan levels, may be in the form of summer camp sites. Consequently it went on growing uninterrupted through PGW-GW and RW-NBPW phases touching ultimately the western active watercourse of the Yamuna river in the last phase. And these days this belt attains highest mark as indicated by 42 closely located small and medium sized villages in this very section as against 69 of the western subsection. This leads to queries as to how these fertile pockets replete with abundant fresh water, pastures and easy game remained barren in terms of habitation in the first major part of the protohistoric past? And how did people manage to inroad into it in the succeeding later half? And more so how is that these pockets are assuming added significance in most recent years? These points are vitally important and would be dealt with in detail a little later.

Frequency of Sites in Rainfall Zones

In the following lines we would analyze our data against the prevailing rainfall trends in this section. And for that matter we must take cognizance of the prevailing trends in the realm of rainfall system, though the official figures being deceptive cannot be projected back in the past as it is, yet, offer good insight. On these lines the area can be classified in several zones. One such zone is comprising of 600-700 mm iso-hyets, which covers only a small part of the north-eastern section of the district. And similarly only some parts of south-eastern section fall within 400-

500 mm rainfall iso-hyets. Major part of the district comes within the parameters of 500-600 mm rainfall zone. This being the largest area enjoys a large number of settlements. This has amply been demonstrated in the table 4.3 (map 10).

The table brings out clearly that the zone with iso-hyets above 600 mm was opted for habitation sometime in the late Harappan times only. But this was not more than a symbolic presence in any respect for the sole site registers mere presence of habitation in these levels unlike the previous phases. But after beginning of this process the number of settlements continued to swell steadily from one in the late Harappan lavels to four and nine in the PGW-GW and RW-NBPW phases respectively. And this was further increased to 23 in this section of the Khadar in modern times. If the number (12) of modern villages of the Bhangar belt

TABLE 4.3

Frequency of Protohistoric and RW-NBPW Sites in the Physiographic and Rainfall Zones

Rainfall Zone	*Physio-graphic Zone*	*EH*	*H*	*LH*	*PGW-GW*	*RW-NBPW*	*Modern Village*
Above 600 mm	Bhangar	00.00	00.00	00.00	00.00	00.00	12 (3.44)
	Khadar	00.00	00.00	01 (1.72)	04 (6.34)	09 (10.91)	18 (5.11)
Above 500 mm	Bhangar	12 (100)	17 (100)	52 (86.66)	52 (82.54)	60 (68.18)	203 (58.33)
	Khadar	00.00	00.00	00.00	04 (6.34)	11 (12.50)	79 (19.09)
Above 400 mm	Bhangar	00.00	00.00	07 (12.07)	03 (4.76)	08 (9.02)	22 (6.89)
	Khadar	00.00	00.00	00.00	00.00	00.00	14 (4.36)
Total		12	17	60	63	88	348

of this rainfall zone is also taken into account the total strength of modern villages comes to 35, which is the highest ever over the times. In the next small area of 400-500 mm rainfall the history of habitation once again began in late Harappan times

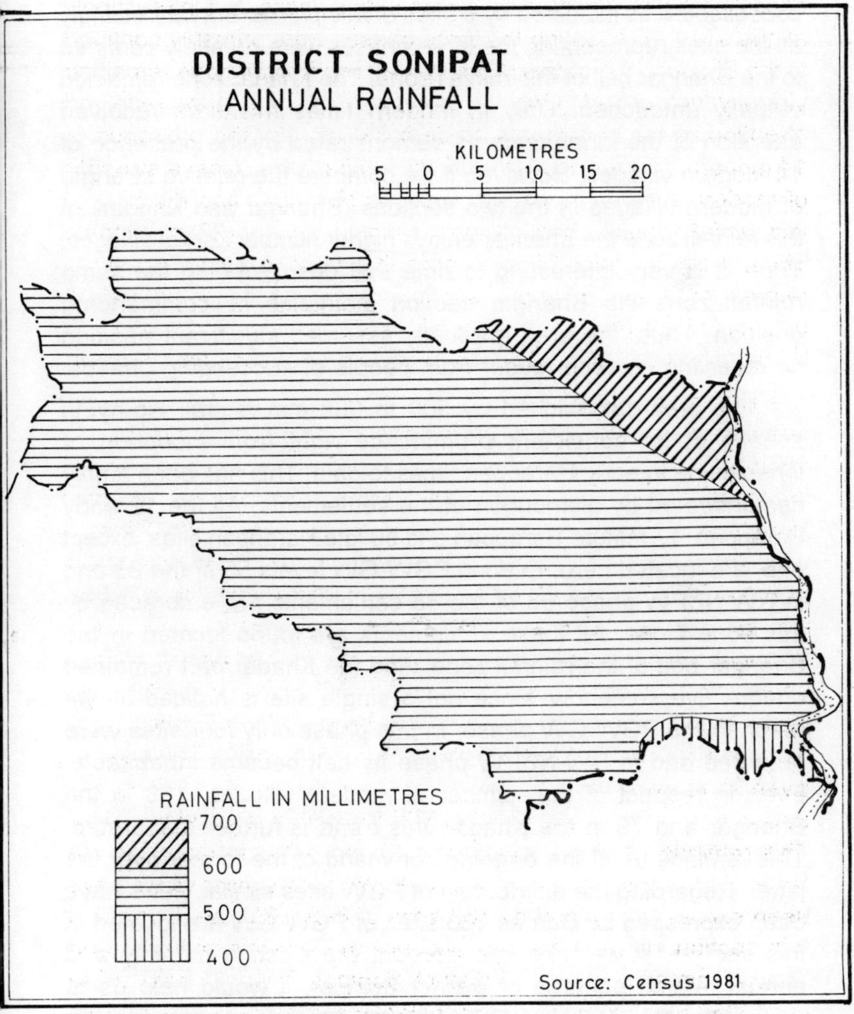
DISTRICT SONIPAT
ANNUAL RAINFALL
KILOMETRES
5 0 5 10 15 20
RAINFALL IN MILLIMETRES
700
600
500
400
Source: Census 1981

Map. 10

with seven settlements, but somehow experienced a reversal in the succeeding PGW-GW phase as only 3 sites represent it. However, in the succeeding RW-NBPW phase once again not only the lost position was regained, but was also further consolidated as indicated by 8 sites of this phase. But interestingly all the sites representing the three phases were primarily confined to the Bhangar belt of this rainfall zone. The Khadar zone remained virtually untouched. Only in modern times this area received attention of the inhabitants as demonstrated by the presence of 14 modern villages. However, if we compare the relative strength of modern villages in the two sections (Bhangar and Khadar) of this rainfall zone the Bhangar enjoys higher number (22) of villages. Thus, it is very interesting to note that despite being the same rainfall zone the Bhangar section maintains its commanding position. Thus, the area gradually assumed significant position by receiving growing favour from people of succeeding phases.

The area sandwiched by 500 to 600 mm rainfall iso-hyets speaks of its consistent prominence right from inception of inhabitation in early Harappan times to date. This has been amply demonstrated by distribution of the settlements. All the 12 early Harappan, 17 mature Harappan and 59 late Harappan sites, except one, are located here. Even in PGW-GW levels 55 of the 63 and in RW-NBPW phase 68 of the 88 settlements are embraced by this zone alone. All these settlements are found located in the Bhangar belt of the rainfall zone. And the Khadar part remained unattractive somehow, since not a single site is noticed till we reach to the PGW-GW phase. In this phase only four sites were recorded and in RW-NBPW phase its belt became inhabitable. Even in respect of the number of modern villages, 198 in the Bhangar and 79 in the Khadar, this trend is further elaborated. This reminds us of the ongoing command of the former over the latter. Regarding the distribution of PGW sites similar views have been expressed by Dutt as 163 sites of PGW-GW are located in this section.[9] If we take into account the existing monthly and annual rainfall patterns of district Sonipat, it would help us to understand the mechanism of constantly attracting people through the times. In this zone the rainfall pattern demonstrates minor fluctuations in normal conditions (fig.4.3-4). And the ongoing rainfall regime provides for nominal to moderate rainfall over the months, specially, after monsoon season, the period for growing

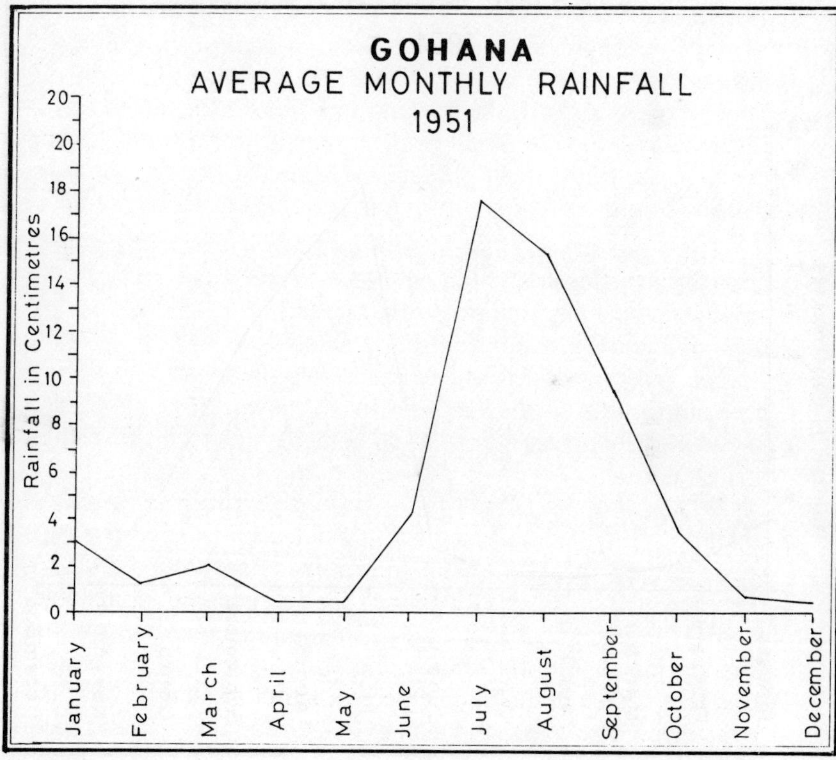
GOHANA
AVERAGE MONTHLY RAINFALL
1951
Rainfall in Centimetres
20
20
18
16
14
12
10
8
6
4
2
0
January
February
March
April
May
June
July
August
September
October
November
December

Fig. 4.3

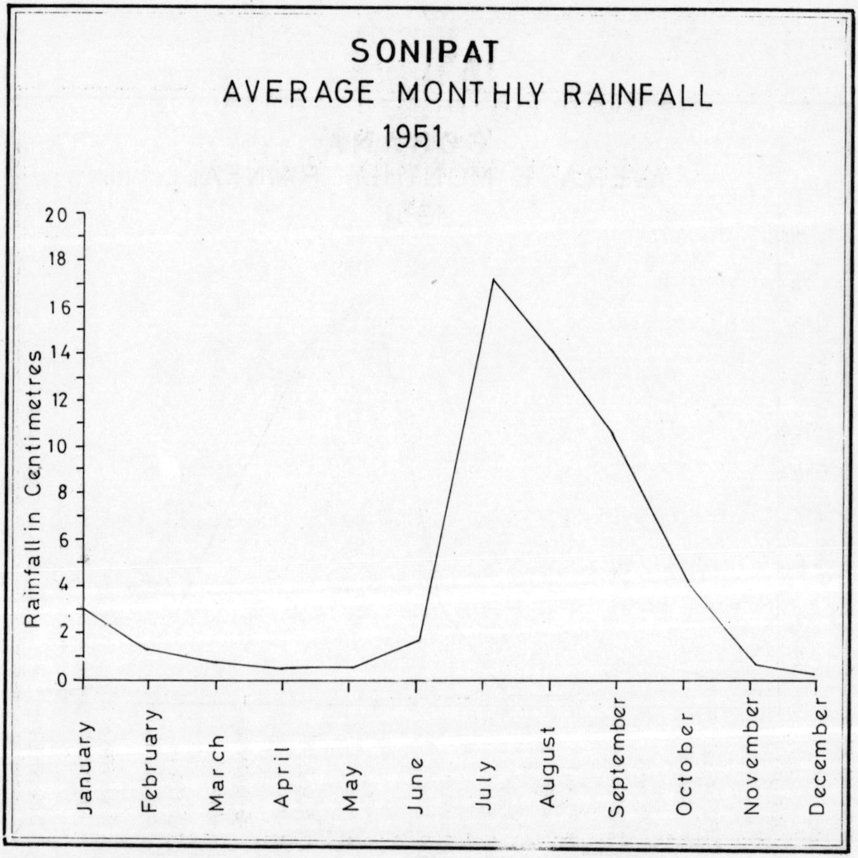
SONIPAT
AVERAGE MONTHLY RAINFALL
1951
Rainfall in Centimetres
0
2
4
6
8
10
12
14
16
18
20
January
February
March
April
May
June
July
August
September
October
November
December

Fig. 4.4

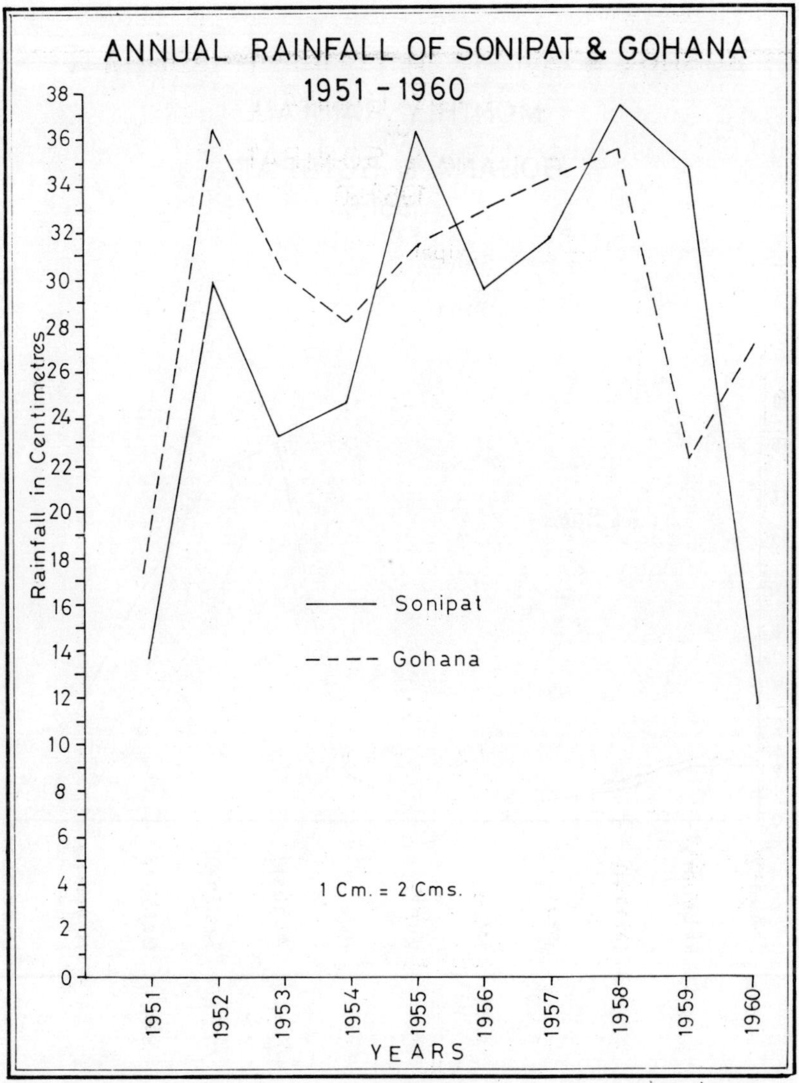
ANNUAL RAINFALL OF SONIPAT & GOHANA
1951 - 1960
Rainfall in Centimetres
38
36
34
32
30
28
26
24
22
20
18
16
14
12
10
8
6
4
2
0
Sonipat
Gohana
1 Cm. = 2 Cms.
1951
1952
1953
1954
1955
1956
1957
1958
1959
1960
YEARS

Fig. 4.5

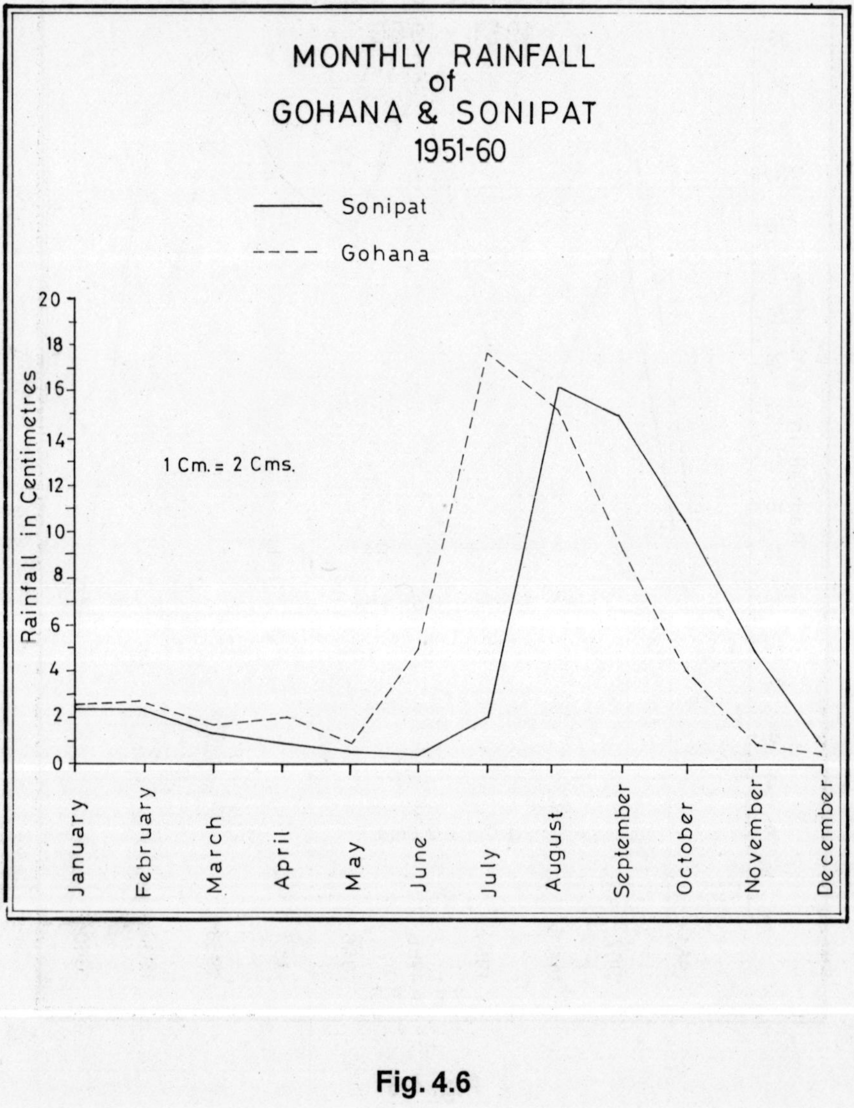
MONTHLY RAINFALL
of
GOHANA & SONIPAT
1951-60
Sonipat
Gohana
Rainfall in Centimetres
20
18
16
14
12
10
8
6
4
2
0
1 Cm. = 2 Cms.
January
February
March
April
May
June
July
August
September
October
November
December

Fig. 4.6

TABLE 4.4

Average Annual Rainfall 1951-60

Station	*1951*	*1952*	*1953*	*1954*	*1955*	*1956*	*1957*	*1958*	*1559*	*1960*
Gohana	13.40 (34.0494)	28.72 (72.97752)	23.56 (59.86596)	22.32 (56.71512)	24.84 (63.11844)	25.76 (65.45616)	26.51 (67.36191)	27.72 (70.43652)	17.32 (44.01012)	20.92 (53.15772)
Sonipat	10.85 (27.56985)	23.45 (59.58645)	17.98 (45.68718)	19.11 (48.55851)	28.22 (71.70702)	22.92 (58.23972)	24.6 (62.5086)	33.68 (85.58088)	26.49 (67.31109)	9.41 (23.91081)
Rohtak	10.68 (27.13788)	24.03 (61.06023)	23.73 (60.2793)	20.97 (53.28477)	21.67 (55.06347)	23.66 (60.12006)	25.53 (64.87173)	25.46 (64.69385)	18.95 (48.15195)	31.82 (80.85462)
Jhajjar	11.91 (30.26331)	21.29 (54.09789)	23.16 (58.84956)	20.2 (51.3282)	23.02 (58.49382)	21.43 (54.45363)	27.55 (70.00455)	27.76 (70.53816)	8.05 (20.45505)	19.28 (48.99048)

Notes: (i) Rainfall figures are in inches, but figures within brackets are in centimetres.

(ii) N.R.—Not received.

Sources. Punjab District Gazetteers (Supplement), 1961.

TABLE 4.5

Monthly Ranifall Average 1951-60

Station	*Jan.*	*Feb.*	*March*	*April*	*May*	*June*	*July*	*Aug.*	*Sept.*	*Oct.*	*Nov.*	*Dec.*	*Total*
Gohana	1.2 (3.0492)	0.5 (1.2705)	0.8 (2.0328)	0.2 (0.5082)	0.2 (0.5082)	1.7 (4.3197)	6.9 (17.5329)	6.0 (15.246)	3.7 (9.4017)	1.4 (3.5574)	0.3 (0.7623)	0.2 (0.5082)	23.1 (58.6871)
Sonipat	1.1 (2.7951)	0.4 (1.0164)	0.3 (0.7623)	0.2 (0.5082)	0.1 (0.2541)	0.7 (1.7787)	6.7 (17.0247)	5.7 (14.4837)	4.2 (10.6722)	1.7 (4.3197)	0.3 (0.7623)	0.2 (0.5082)	21.6 (54.8856)
Rohtak	1.2 (3.0495)	0.6 (1.5246)	0.7 (1.7787)	0.3 (0.7623)	0.3 (0.7623)	1.2 (3.0492)	7.5 (19.0575)	5.7 (14.4837)	3.5 (8.8937)	1.2 (3.0492)	0.2 (0.5082)	0.2 (0.5082)	22.6 (57.4266)
Jhajjar	0.6 (1.5246)	0.5 (1.2705)	0.4 (1.0164)	0.2 (0.5082)	0.3 (0.7623)	1.0 (2.541)	6.7 (17.0247)	6.2 (15.7542)	2.8 (7.1148)	1.4 (3.5574)	0.2 (0.5082)	0.1 (0.2541)	20.3 (51.5823)

TABLE 4.6 : GOHANA

Monthly Ranifall 1951-60

Month	1951		1952		1953		1954		1955	
	Rainfall	Rainy Day	Rainfall	Rainy Day	Rainfall	Rainy Day	Rainfall	Rainy Day	Rainfall	Rainy Day
January	0.47	1	0.59	2	2.78	5	1.56	3	1.79	5
February	0.00	–	0.75	4	0.00	–	3.58	6	0.04	–
March	0.70	3	0.70	1	0.00	–	1.43	1	0.08	–
April	0.47	1	0.12	1	0.03	–	0.00	–	0.32	1
May	0.05	–	0.86	1	0.14	–	0.24	1	0.00	–
June	2.00	3	4.38	6	4.33	4	0.29	1	0.70	1
July	0.50	2	7.80	13	8.43	8	8.04	9	2.69	5
August	6.14	10	13.49	13	5.69	9	2.28	2	8.05	9
September	2.12	4	0.00	–	1.84	3	3.22	9	6.72	8
October	0.00	–	0.00	–	0.00	–	1.68	2	4.42	6
November	0.95	1	0.00	–	0.00	–	0.00	–	0.00	–
December	0.00	–	0.03	–	0.32	1	0.00	–	0.03	–
Total	13.40 (34.0)	25	28.72 (73.0)	41	23.56 (59.8)	30	22.32 (56.7)	34	24.84 (63.1)	35

Notes (i) Rainfall figures are in inches, but figures within brackets are in centimetres.

(ii) N.R.—Not received.

Sources: Punjab District Gazetteers (Supplement), 1961.

TABLE 4.6 : GOHANA

Monthly Ranifall 1951-60

Month	1956		1957		1958		1959		1960		
	Rainfall	Rainy Day	Rainfall	Rainy Day	Rainfall	Rainy Day	Rainfall	Rainy Day	Rainfall	Rainy Day	Average
January	0.56	2	1.98	4	0.15	1	1.50	2	0.40	1	1.2
February	0.50	1	0.00	–	N.R	N.R	0.21	1	0.00	–	0.5
March	0.83	2	1.49	3	0.31	2	0.40	1	2.20	2	0.8
April	0.12	1	0.00	–	N.R	N.R	0.00	–	1.19	2	0.2
May	0.00	–	0.00	–	0.40	1	0.06	–	0.00	–	0.2
June	1.03	2	1.32	4	0.65	3	0.10	1	2.37	3	1.7
July	11.59	9	7.66	8	7.70	9	4.14	3	11.07	10	6.9
August	4.54	8	3.33	4	4.50	11	7.84	8	3.69	4	6.0
September	1.15	3	7.60	6	12.11	17	1.85	3	0.00	–	3.7
October	4.74	7	2.45	2	0.50	1	0.00	–	0.00	–	1.4
November	0.70	1	0.18	1	0.00	–	1.22	1	0.00	–	0.3
December	0.00	–	0.50	2	1.40	1	0.00	–	0.00	–	0.2
Total	25.76 (65.4)	36	26.51 (67.3)	34	27.72 (70.4)	46	17.32 (44.0)	20	20.92 (53.1)	22	23.1 (58.7)

Notes: (i) Rainfall figures are in inches, but figures within brackets are in centimetres.

(ii) N.R.—Not received.

Sources: Punjab District Gazetteers (Supplement), 1961.

TABLE 4.7 : SONIPAT

Monthly Ranifall 1951-55

Month	*1951*		*1952*		*1953*		*1954*		*1955*	
	Rainfall	*Rainy Day*	*Rainfall*	*Rainy Day*	*Rainfall*	*Rainy Day*	*Rainfall*	*Rainy Day*	*Rainfall*	*Rainy Day*
January	0.64	1	0.21	N.R	1.29	2	0.86	2	1.22	5
February	0.00	–	0.80	"	0.00	–	3.03	4	0.22	1
March	0.52	1	0.45	"	0.00	–	0.40	1	0.07	–
April	0.40	1	0.22	"	0.00	–	0.04	–	1.55	2
May	0.00	–	0.59	"	0.00	–	0.00	–	0.04	–
June	0.09	2	1.43	"	0.90	2	0.29	2	1.00	1
July	0.81	1	9.15	"	10.35	10	6.82	7	3.67	5
August	5.63	9	10.42	"	4.90	7	0.80	2	6.34	8
September	1.26	4	0.00	"	0.24	1	3.59	7	10.99	7
October	0.00	–	0.00	"	0.00	–	3.28	2	3.12	6
November	0.50	2	0.00	"	0.00	–	0.00	–	0.00	–
December	0.00	–	0.18	N.R	0.30	1	0.00	–	0.00	–
Total	10.85 (27.6)	21	23.45 (59.6)	N.R	17.98 (45.7)	23	19.11 (48.5)	27	28.22 (71.7)	35

Notes: (i) Rainfall figures are in inches, but figures within brackets are in centimetres.

(ii) N.R.—Not received.

Sources: Punjab District Gazetteers (Supplement), 1961.

TABLE 4.7: SONIPAT

Monthly Ranifall 1956-60

Month	*1956*		*1957*		*1958*		*1959*		*1960*		
	Rainfall	*Rainy Day*	*Rainfall*	*Rainy Day*	*Rainfall*	*Rainy Day*	*Rainfall*	*Rainy Day*	*Rainfall*	*Rainy Day*	*Average*
January	1.20	1	1.98	5	0.15	1	1.53	2	1.38	3	1.1
February	0.00	–	0.00	–	N.R	N.R	0.00	–	0.00	–	0.4
March	0.29	1	0.34	1	N.R	N.R	0.00	–	1.35	3	0.3
April	0.11	1	0.00	–	N.R	N.R	0.00	–	0.00	–	0.2
May	0.00	–	0.08	–	0.00	–	0.00	–	0.00	–	0.1
June	0.41	–	0.00	–	0.00	–	1.96	3	0.00	–	0.7
July	3.66	7	12.06	7	7.52	8	6.45	7	6.68	9	6.7
August	5.61	7	3.05	3	11.61	7	8.30	8	0.00	–	5.7
September	0.38	2	6.27	5	12.40	8	6.87	3	0.00	–	4.2
October	10.09	6	N.R	N.R	0.16	1	0.63	1	0.00	–	1.7
November	1.17	1	0.18	1	0.00	–	0.75	2	0.00	–	0.3
December	0.00	–	0.10	1	1.84	1	0.00	–	0.00	–	0.2
Total	22.92 (58.2)	27	24.06 (61.1)	23	33.68 (85.5)	26	26.49 (67.3)	26	9.41 (23.9)	15	26.6 (54.9)

Notes (i) Rainfall figures are in inches, but figures within brackets are in centimetres.

(ii) N.R.—Not received.

'barani' rabi crops. This is further explained by figures 4.5-6 and Tables 4.4-7. Thus, the area being capable of producing nominal to moderate *kharif* as well as *rabi* crops promised people the availability of subsistence for their existence. Similarly the area being away from any river and with flood risk free elevated zone provided them safer habitation as well as arable locales to continue over there. Whereas the Khadar being on the one hand the Yamuna flood plain and belonging to the same rainfall zone on the other was bound to receive added amount of water year-round in general and during rainy season in particular. This was bound to aggravate the existing flood threats and thereby risk to life and crops. With the result the abundant water was perceived as a deterrent rather than any stimuli by the people to prefer it over the Bhangar for habitation. Here the transition of rain iso-hyets from one physiographic zone to the other affects the approach of the people. However, it is not the rain alone, which governs the choice of site selection. It worked effectively in combination of other forces. Whenever there was a harmony among the various operative variables their combined strength started things moving.

TABLE 4.8

Frequency of Protohistoric and RW-NBPW Sites in Subsoil Water-level Zones

Water-level	*EH*	*H*	*LH*	*PGW-GW*	*RW-NBPW*	*Modern*
1-2 m. Khadar	—	—	01 (1.66%)	08 (11.11%)	23 (26.13%)	111
2-4 m. Eastern Bhangar	03 (25%)	03 (17.65%)	19 (31.44%)	28 (44.44%)	42 (47.72%)	185
4-6 m Western Bhangar	09 (75%)	14 (82.33%)	40 (66.66%)	27 (42.85%)	23 (26.13%)	52
Total	12	17	60	63	88	348

Frequency of Sites in Subsoil Water Level Zones

The table 4.8 indicates that the area with 1-2 metre deep water-level did not find favour with people of the first two phases of Harappan civilization as these draw blank in regard to their settlements. And in the last phase—late Harappan also only one site is recorded. Since the area conforms to the Khadar with high watertable and consequential hardships of waterlogging, swampy conditions and thick vegetal cover infested with wild animals and insects would have restricted their movements in this zone in absence of matching technical know-how and organization. The next area with subsoil waterlevel ranging between 2-4 metre was slightly better placed in context of habitation sites enjoying 3 each in early and mature Harappan, 19 in late Harappan, 28 in PGW-GW and 42 in RW-NBPW levels as this would have posed fewer problems compared to the first zone in the east. Interestingly the area with 4-6 metre deep subsoil water-table appeared most conducive to the people in the past as it contains 9 of the 12 early Harappan, 14 of the 17 mature Harappan, 40 of the 60 late Harappan, 27 of the 63 PGW-GW and 23 of the 88 RW-NBPW sites. This allocation denotes constant importance of this zone in all the stages of Harappa Civilization. Afterwards a gradually growing shift from this zone to 2-4 metre water depth zone is attested to. And the fallout of this eastward advancing movement is also felt in next zone—eastern with 1-2 metre water-table. This tendency is still visible in case of modern villages. Here the first and second zones support 111 and 185 of the 348 villages respectively. Only 52 villages are now harboured by once the most conducive appearing western subzone in Harappan times.

Frequency of Sites in Groundwater Zones

Let us now put our data to another mode of analysis based on nature of groundwater quality, which is given in table 4.9.

TABLE 4.9
Frequency of Protohistoric and RW-NBPW Sites in Ground Water Zones

Ground water	*EH*	*H*	*LH*	*PGW-GW*	*RW-NBPW*	*Modern*
Fresh Water	—	—	02 (3.33%)	08 (11.11%)	21 (23.86%)	111 (31.90%)
Fresh water in Shallow Aquifers	—	—	02 (3.33%)	12 (19.04%)	26 (29.54%)	48 (13.79%)
Fresh water in Deeper Aquifers	—	—	—	—	—	—
Brackish to Saline water	12 (100%)	17 (100%)	56 (93.33%)	43 (68.25%)	41 (46.60%)	188 (54.31%)
Total	12	17	60	63	88	348

This table displays the whole range of explored settlements and explains that not a single site of early and mature Harappan phases was found in the fresh water zones. These people were fond of the brackish to saline water zone as all the settlements of these periods were located here. This zone remained conducive for inhabitants, even in succeeding levels. This is marked by 56 of the 60 late Harappan, 43 of the 63 PGW-GW and 41 of the 88 RW-NBPW sites. So far as the present day villages are concerned 188 of the 348 are supported by this zone.

Regarding first two categories of fresh water zones virgin soil seems to have been broken afresh in late Harappan times. Gradually, as the time passed on, the number of PGW-GW and RW-NBPW sites is visibly growing from 4 in late Harappan times to 20 and 43 in PGW-GW and RW-NBPW phases respectively. This is further enhanced in modern times to 159 of the 348 modern villages. These points signify the orientation of the gradual dissemination from west to east of farming communities in this section. In the earlier phases it was perhaps not feasible to penetrate in the eastern pockets. Therefore, their centre of frequent interaction was restricted to the brackish to saline water zone conforming to the Bhangar in the west (map 4).

Frequency of Sites in Different Cropping Systems

The table 4.10 shows frequency of sites in different cropping zones. Before dwelling on niceties of classification of settlement data, it would be appropriate to make some observations in this connection. First the present cropping pattern is fast changing under the influence of modern scientific developments and researches. Farmers adapt themselves quickly according to the changing environment. They adopt and experiment new seeds, fertilizers and techniques. Their emphasis is on high yielding crops and intensive farming. Whereas the similar frequent experiments were not even imaginable in the past, especially till the beginning of green revolution.Their cropping system was perhaps largely based on traditional knowledge of crops, information of probable trends of ecology and techniques of subsistence. It was a very slow process, at times with imperceptible changes. It is therefore, impractical to depend wholly on current cropping patterns. This problem is further increased especially in absence of direct evidence of past crops from the area under discussion. We have received information from the contemporary excavated settlements from the adjoining areas and that of recent past, especially of pregreen revolution period, crop systems of the area to workout a tentative cropping system of yester-years. Such a combined body of information would perhaps reduce the degree of uncertainty to a large extent. The cropping pattern adopted would be a combination of the two aforesaid trends. This analysis is of course subject to fresh enquiries.

The aforesaid table 4.10 shows classification of sites on the basis of rice-maize and millet-jowar crops of *kharif* season. In *rabi* season crops like wheat, barley and gram have been selected as being dominant representative of the two periods—past and present. This table highlights the following vitally significant points. There is no site of the first two Harappan phases in the rice and maize zone, it was first ever opted by the late Harappan people as indicated by the presence of their two sites. Following this more people were attracted to make 8 and 23 sites of relatively large size in PGW-GW and RW-NBPW phases respectively. Thus, in succeeding phases increasing importance was attached to this section. This tendency is further exhibited by the presence of 111 modern village settlements.

TABLE 4.10

Frequency of Protohistoric and RW-NBPW Sites in Cropping Systems

Crop-System	*Crops*	*EH*	*H*	*LH*	*PGW-GW*	*RW-NBPW*	*Modern*
Kharif	Rice+ Maize	—	—	02 (3.45%)	08 (11.11%)	23 (26.16%)	111 (31.90%)
	Millet+ Jowar	12 (100%)	17 (100%)	58 (96.66%)	55 (88.89%)	65 (73.84%)	237 (68.10%)
Rabi	Wheat+ Gram	12 (100%)	17 (100%)	58 (96.66%)	63 (100%)	88 (100%)	348 (100%)
	Barley+ Gram	12 (100%)	17 (100%)	58 (96.66%)	55 (88.89%)	65 (73.84%)	237 (68.10%)

Regarding analysis of sites in context of another combination of millet-jowar crops of *kharif* season, the area producing these crops conforms with the upland dry Bhangar. Interestingly sites belonging to all the three stages of Harappa, barring two of the last phase, enjoyed the millet-jowar producing zone. This very zone, however, constitutes primarily gram-barley producing belt also. It would, therefore, not be unfair to conclude that the Harappan people keeping in view the prevailing ecological and techno-organizational conditions fully appreciated the drought resistant nature of these crop varieties, which were best suited not only to the physical features, but also the existing effective strength of the people equipped with simple and limited conventional know-how. This practice was very popular and successful until very recent past in this area, i.e., before beginning of green revolution. In modern times also 237 villages out of 348 are receiving their subsistence from this zone alone. But soon after the Indus phase, i.e., in the PGW-GW and RW-NBPW phases, an ever increasing tendency to go for the rice-maize zone is noticeable. These two phases represent 8 and 23 settlements respectively in this section. The zone so attractive in Harappan times began drifting away its magnetic influence over the inhabitants. A similar tendency is also quite visible among the modern people of 111 villages of the Khadar zone. This once again testifies the 'gradual advancement' of farming communities from the west to the east. Such a situation is neither spontaneous nor stimulated by any sole factor. This seems to have been a result of a combination of varied, but interacting forces. These are to be taken up in the following discussion.

The discussion on locational position of ancient settlements in the preceding pages brings out certain very interesting features as summarized below. A large number of settlements, if not all, are found located in the Bhangar region. The Khadar zone invited a few people and that too in subsequent stages beginning with late Harappan phase onwards. So far as the physical properties of the two zones are concerned the former is devoid of any perennial river, good quality of sweet subsoil water in abundance within reach of the people, fertile alluvial soil with a natural refreshing mechanism through annual inundation, strong possibility of surface water in ox-bow lakes, rich environments for pasture grounds and games, etc., unlike the low-lying Khadar flood plain of the Yamuna river in the east. The general contention that the areas with easily available water and better soils were the first preference of the protohistoric people than the hard and dry area[10] does not hold out any longer in view of the contrary findings in our own area. It is very curious and needs explanation as to how the people in the ancient past preferred to the Bhangar over the Khadar? And how subsequently emphasis shifted growingly towards the Khadar in course of time? The explanation of these queries seem to lie with the very inherent contrast in the physical and ecological conditions of the two zones, the past and present subsistence practices, the difference in population size, changing technological levels, etc. The Khadar of course, is very rich in water—both surface and underground, land-forms and vegetation than the Bhangar. These very salient features appear to have proved deterent in the ancient past to making new settlements instead of any attraction. Because this was fraught with many dangers to human lives whether be it flood hazards as generally rivers of greater size are more of a danger than attraction to peoples with limited technical expertize,[11] menace of wild animals, inpregnable growth of natural grasses and vegetation, waterlogging conditions and scarcity of suitable topographic variations to provide frequent reasonably raised grounds. In, as we know, the ancient past people were not equipped with suitable tools, techniques and human organization to harness such natural environment, control Yamuna flood fury, clear dense vegetal cover, exploit abundant natural fresh water to their benefit. But as we witness some changes in the then prevailing environment of technology with the introduction of iron, and level of human organization with the steady growth in populations, especially in second half of late Harappan and the beginning of PGW phases, the ongoing locational trend started experiencing a shift from the west to the east leading to growing concentration in the hitherto inhospitable and intractable Khadar zone. This becomes more obvious in the following phase of RW-NBPW in which the number of settlements increases to 19 from 8 of the PGW-GW phase. And if we consider the number of 111 modern villages the trend becomes crystal clear. This shows the overwhelming role and impact of changes in technology and organization.[12] These two factors help promote capacity and confidence to face, if not to overcome, intractable looking natural environments.

While just contrary to the Khadar zone there were no such dangers in the Bhangar. Neither there was any perennial river nor any high water-table of fresh water supporting growth of lush green cover of natural vegetation harbouring dangerous wild animals densely. Nature of course, has provided a working mechanism for surface-water in between through recurrent occurrence of depressions of varying sizes and proportions. These depressions might have been of greater size in the past and must have assumed significant proportions especially in rainy season, though rainfall system in itself remains erratic, precarious and undependable, yet, usually it is of moderate nature and spread over fairly in space and time as indicated by table 4.4-7 and figures 4.3-6. Similarly there was a very rich combination of fertile rising loose sandy patches with less vegetal cover, fairly flat loam and clayey low-soils, suited to the then prevailing ecological, technological, organizational conditions and subsistence practices to meet the requirements of the people. The existence of a large number of modern villages and their conditions about 50 years back, in some cases even now, can help provide further insight and consequent affirmation of such an observation. The villagers, as a rule, did not have any access to natural fresh water (a large number of them do not have even now), since the quality of sub-soil water is brackish to saline. They used to depend on surface rain-water available either in natural or man-made ponds (this practice is still in use in parts of Ganganagar district) or fresh subsoil water available in shallow aquifers through wells dug on the periphery of these very ponds usually (see figs. 5.54-56). As these ponds used to receive rain-water, there was a fair possibility to maintain sufficient amount of water for most part of the year in normal rainfall conditions. The continually stagnant water in these ponds would effect transformation in the nature of subsoil water in varying degrees depending upon the intensity and time span of percolation. Thus, the inflow, level and quality of water were managed in these wells to serve water needs of the people with a certain amount of stability. In this connection some deep circular pits exposed outside the habitational area at Atranjikhera may be mentioned as supportive evidence indicating existence of *kuccha* wells for water purposes.[13] If this mechanism could form the main lifeline in the recent past for existence of life in this area can we not visualize a similar pattern in the ancient past? This hypothesis gains strength especially, when most of the ancient settlements are found located along such natural depressions irrespective of their size. Similar observation was made by Lal regarding location of sites in the Ganga–Yamuna

Doab especially away from rivers.[14] Moreover despite the threats of droughts the Bhangar soils were less formidable than the flood plains of the Yamuna in Harappan times with simple copper technology and even in subsequent PGW-GW levels with limited and simple tools of iron. Hence, in the given situation this sort of natural combination offered ample avenues for bare subsistence activities. And that is what we witness in this zone.

Factors Promoting Uneven Settlement Patterns

We have already discussed the spread of settlements. Settlements being an embodiment of complex system produced by a variety of interacting, but complimentary factors. Here we have first to understand the constituent variables and secondly to analyze their relative role to influence the choice of making settlements. As we have already underlined, during the period under review, a constant increase in the number of settlements. Likewise more and more prestine areas came under influence during their dissemination with eastward thrust from the early Harappan to RW-NBPW levels. This spread marks out different areas of varying concentrations from the west to the east until it embraced the Yamuna flood plain. Therefore, it is imperative to identify the forces promoting such settlement patterns.

In the first two phases of the Harappan civilization the area situated to the west of Gohana, the western subzone of the Bhangar, was the centre of concentration of settlements of these two phases. During this first stage of settlements the people were equipped with copper technology. Its circulation was limited in the early Harappan levels, but this experienced a reasonable increase during the succeeding mature Harappan times. Similarly there was also some improvement in the tools and their techniques. Nevertheless it still could not divorce of its all the inherent limitations for the metal in itself being soft and the technology employed being simple and conventional.[15] And in the face of the tough natural conditions of this area it was not to be quite useful beyond a point. These two constraints impaired the working capacity of the people. In order to come out of this eclipsed potential, if not to transcend it completely, at least to maintain their existence more hands were needed to improve their working capacity. However, this was also subject to the three very crucial aspects, first is the number of working persons available at a settlement and at other settlements around, second is the level of mutual cooperation either generated by a common threat perception or by a mature decision or by both and third is the clarity of the objective of the people. Obviously average human beings aspire to rise above their present level of living. And for that matter they not only employ the extant resources judiciously, but also try to explore new ways and means to achieve that objective. Likewise the early Harappan people would have aspired to enhance their working capacity, the operative potential, but the small-size and number of settlements do not promise any extra hand to increase their work-force. And in this situation, even their well coordinated efforts were not suffice to effect any substantial change in their extant working capacity. With the result the people were not capable enough to enforce any tangible innovations beyond protecting their existence and perhaps that was their prime objective in that situation. However, in the mature Harappan levels the enlarged size of settlements do suggest an increase in the population, which not only warrant additional hands, but also greater scope for better coordination. This was also aided by the growing knowledge about the behaviour and potential of the natural forces owing to their constant interaction over the long period. This increased the operative potential of the people to negotiate with their environment with greater vigour and wisdom. And this is reflected in both the greater level of production and the enlarged area of resource exploitation than that of the early Harappan times. Habitations were firmly established at the preceding ones as well as at the new locations, but strictly in the subzone of the Bhangar. This was further elaborated as far as the number of settlements is concerned in the late Harappan levels.

In the second area of concentration, the eastern subzone of the Bhangar, between Gohana and Sonipat towns, a number of habitations were made for the first time during late Harappan levels. However, during this phase we do not have any evidence to perceive any change in the realm of natural conditions as these fresh pockets do bear in as much as the natural features are concerned a striking similarity with those of the western sub-zone of the Bhangar. As regards technological know-how copper technology was still in vogue, which registered an increase in the number of tools, especially with the introduction of copper Hoards. Certain instances of copper objects do exhibit some improvement over the preceding ones. This is further strengthened by the exodous of a large number of persons from the urban areas to the hinterlands, in this area as well, betraying dissemination of certain production tools and ideas along with them. Besides the very increase in the number of settlements is suggestive of further additions in the size of population as a result of internal growth and immigration, if 37 and 19 sites in the west and east sub-zones respectively are any indication. This growth of population supplied fresh hands to strengthen their manpower. This increase in population would also have called for more foodgrains. And this could have been met with by opening up new fields around to exploit the prestine lands. In addition to this we have also evidence of introduction of new crops such as millet and sorghum, which provided them with more opportunities to raise crops and thereby food. Moreover over the period people became more enlightened to assess the behaviour and the potential of the natural forces. Therefore, in the given situation they employed their operative potential more realistically and managed to survive in the hard conditions of this hitherto untraversed subzone. In the succeeding PGW-GW levels two very important changes took place, one is the introduction of iron-tool technology, though on a limited scale and the second is paddy cultivation by way of transplantation. These two innovations were certainly capable of improving their operative potential. This was coupled with further increase in population and thereby the manpower, if the number and size of settlements (22) is any pointer. Thus, people armed with additional information about the behaviour and potential of the natural resources, new crops to be grown, new tools and techniques to be employed and greater manpower to fall back upon, were surely better equipped to negotiate with their natural surroundings than ever before. The combine of these forces explains the growing operative potential and the consequent better concentration of late Harappan and PGW-GW settlements in the eastern sub-zone of Bhangar.

The Western Margins of the Yamuna Khadar forms the dividing line of the Bhangar in the west and the Khadar in the east. Being the meeting point of the two contrasting physiographic zones it combined the characteristic features of the two zones and with the result was certainly better than of the two main zones. Nevertheless it also suffered from certain constraints especially in terms of the expanse of this zone for it basically constituted a narrow serpentive strip promising subsistence opportunities to a limited number of people with the help of the then available copper tools, crop systems and limited manpower as indicated by a few small late Harappan settlements (3). With the addition in tool technology, crops and manpower the existing operational potential experienced improvement and with the result a greater number of PGW-GW settlements (10) of medium and bigger size came into being as has been observed earlier. Settlements like Gumar, Agwanpur, Sandhal Khurd, Jawahari–I, Sonipat, Harsana Kalan–I and Akbarpur Barota, perhaps pertain to the second phase of the PGW,[16] suggest that. Later in the early historical period additional improvement was experienced in the iron-tool technology as suggested by the occurrence of iron ploughshares and reference of heavy ploughs drawn by several oxen. Instances of dung manuring points to further increase in the agricultural knowledge. Relatively the higher number (18) and the bigger size of settlements along the Western Margins speak of increased man-power and higher level of mutual coordination. And the greater frequency of cultural artifacts explain the higher level of production. The knowledge of coins, punch maked, might have accelerated the pace of interaction during this period. What we witness here along the Western Margins was a product of the combined effect of the change in some of the operative variables from the late Harappan down to early historical levels. This also underlines that a better combine of natural and social forces could produce better results.

The last area of better concentration is the Khadar zone, which attains this status only in the early historical levels with RW-NBPW, though it had experienced human presence in late Harappan levels with one site and still better human presence in PGW-GW, especial GW levels with eight settlements. However, even in the early historical levels this number of habitations was not really high in view of the vast expanse and the biodiversity of this zone. But this very number of settlements in this vast zone is very important to betray the limitations of the operative potential in the face of the dangerous zone fraught with many problems to life and property though apparently a conducive looking zone.

And consequently people were not independent enough to make abodes wherever they liked to settle down. However, this very zone was more unconducive during the entire Harappan times. And this formidability was reduced a little during the PGW-GW settlements with an improvement in the operative potential as we have noticed earlier. This was further improved in the RW-NBPW levels. And with the further transformation in the operative potential in the modern times this zone became the most thickly populated zone. What we witness in this zone is the upcoming settlements very close to the active course of the river Yamuna irrespective of its inherent flood hazards. Had water either alone or in combination with good natural conditions been the prime stimulant to make settlements and their nucleus or the vice versa, then this zone alone should have been preferred to over the dry Bhangar right from the beginning of settlement history here. Whereas the Bhangar is just contrary to the Khadar in so far as the natural conditions are concerned and nevertheless remained the centre of human activities during the period under discussion. And it is also true that the operative potential was no different than those of the Khadar. But, even then how was that the Bhangar enjoyed more confidence of the people? In fact the available operative potential was effective in the Bhangar zone because of the very presence of frequent loose and elevated soils composed of aeolian sand with sparse vegetation cover. Such soils can easily be worked with simple tools, minimum tillings and labour to prepare them as seedbeds. And what we experience in regard to their site selection is imbibed with these pragmatic considerations as they selected aeolian sandy loose and elevated pockets[17] on an average. These are also located along some natural depressions akin to that observed in Saurashtra.[18] They grew crops, as mentioned earlier, on such soils with the help of extensive 'barani' cultivation system. Hence, the operative potential was most successful in such areas alone and with the result the Bhangar was not as dangerous as was the Khadar. Besides the varied denomination of settlements in both the zones was the product of uneven dissemination of the forces of production. Wherever the combine of these forces was better the process of development is observed in accordance with. Only occurrence of better locales with suitable natural resources alone was not enough[19], but, whenever it was coupled with at least some, if not all, of the constituents of the operative potential proved significant in the distribution of settlements. Therefore, it is neither the natural resources alone nor any other single constituent of the forces of production, which can be considered supreme to produce desired results. It is rather a combine of the natural and social forces to offer a potential force to realize the social objectives. And the degree of realization of their social needs depends upon the level of harmony among the forces of production.

Rank-size Frequency of Sites and Demographic Analysis

The rank-size frequency analysis forms an integral part of settlement studies. The ranking depends upon internal differentials of sites, their varied sizes, functions carried out on them and relationship with the model or representative settlements in specific areas. Factors like the absence of excavated settlements to unfold internal differentiation, the multiculture nature and disturbed condition of settlements impede a more scientific and realistic ranking of all the recorded settlements. In some cases the number of sites to be analyzed is small as well because such inherent constraints of this exercise. However, this attempt would cover as many settlements as possible to lend it authenticity and emphasis would be on both the single and major culture/representative sites to determine the rank-size category. Such inferences tend to be highly tentative, nevertheless would offer further insight about life of the people.

Before we proceed to the main discussion it would not be out of place to state at the outset that the whole range of collected sites with varied denominations would be divided into seven categories. Each category representing a definite size in respect of area and the probable number of inhabitants. This classification, despite all limitations, is opted for our own convenience to cover the wide range of settlement size and to regain, if at all possible, the past processes of development of settlement hierarchy and system. According to this methodology some selected sites are presumed representative of the respective cultures, which could help understand, their demographic patterns in respect of both the individual settlements and the overall size of population thereby.

TABLE 4.11

Early Harappan Sites: Rank-size Frequency*

Rank Size	*Area of Site (In. sq. m.)*	*Number of Single/ Major Culture Sites*	*Population Size:20 Per 1000 sq. m.*
A	Approximate 2000 sq. m.	7(58.3%)	40
B	5000 sq. m.	—	100
C	10000 sq. m.	—	200
D	15000 sq. m.	—	300
E	20000 sq. m.	—	400
F	30000 sq. m.	—	600
G	Above 30000 sq. m.	—	600

* See Appendix-II for single or major culture sites of this period.

Rank-size Frequency of Early Harappan Sites

Of the 12 sites of this culture only 7 are picked up for rank-size analysis. These have been selected because either they are single culture sites or present a fairly large circulation of characteristic ceramic traits or are found in a relatively better physical condition than the others. Settlements like Bahgru, Silana, Rukhi–I, Rindhana–II and Nuran Khera–III have been omitted as these lack in the aforesaid features.

The analysis in table 4.11 points that all the sites fall in category–A encompassing approximately 2000 square metre area and promising to accommodate about 40 persons as per the working arrangement made keeping in view all essential additions and omissions in this regard. This shows that 58.3 per cent settlements are very small in size. None of them even could attain the next category-B of 5000 square metre accommodating around 100 persons. On this basis, the rest of the early Harappan settlements also seem to belong to the same category if the low frequency of cultural artifacts from these settlements is any pointer. And thus, the analysis of these and the cultural artifacts contained by them speak of common practices in vogue. This also underlines the precarious conditions on an average during this phase.

TABLE 4.12

Mature Harappan Sites: Rank-size Frequency*

Rank Size	*Area of Site (In. sq. m.)*	*Number of Single/ Major Culture Sites*	*Population Size 20 Per 1000 sq. m.*
A	Approximate 2000	01 (6%)	40
B	5000	04 (23.1%)	100
C	10000	02 (12.4%)	200
D	15000	01 (6%)	300
E	20000	01 (6%)	400
F	30000	—	600
G	Above 30000	—	600

* See Appendix-II for single or major culture sites.

TABLE 4.13

Late Harappan Sites: Rank-size Frequency*

Rank Size	*Area of Site (in Sq. m.)*	*Number of Single or Major Culture Sites*	*Population Size 20 Per 1000 sq. m.*
A	Approximate 2000	50 (10.2%)	40
B	5000	30 (61.2%)	100
C	10000	03 (12.2%)	200
D	15000	—	300
E	20000	—	400
F	30000	—	600
G	Above 30000	—	600

* See Appendix-II for single or major culture sites.

Rank-size Frequency of Mature Harappan Sites

This analysis of mature Harappan sites as given in the table 4.12 includes 9 of the 17 mature Harappan sites. And sites like Ahulana–I, Chhapara, Gangesar–I, Ghadwal–II, Kahalapa–II, Rindhana–II, and Nuran Khera–I, etc., have been excluded for absence of the above qualifying features. However, the distributional pattern of the selected sites represents the first five categories from A to E except the last two—F and G, though their respective numerical strength of settlements is as low as 1, 4, 2, 1, and 1 from A to E rank-size group supporting their corresponding number of persons per settlement from about 40 to 100, 200, 300 and 400 respectively. This distribution further highlights that unlike the previous phase the number of small sites (1) representing category–A is very meagre. The average size settlements in this phase seem belong to the next higher category–B with 4 sites (23.1 per cent). The next higher category–C is represented by 2 (12.4 per cent) sites. This aggregation trend does not stop until embraced categories D and E represented by one settlement each. This amply demonstrates that the period experienced a tremendous thrust to gain size, strength and stability. Nevertheless it remained short of achieving the other two higher categories F and G, which further suggests that none of the mature Harappan sites could join the class of Balu (60,327), Mitathal (41,200), Siswal (47,100), etc. with their respective area in square metre.[20] Therefore, the Harappan sites of this district primarily remained village settlements harbouring people of Siswal stock like Siswal B.[21]

Rank-size Frequency of Late Harappan Sites

In connection with the rank-size analysis of the late Harappan settlements 38 out of 60 sites have been taken up as shown in table 4.13, which shows that only a small number of 5 (10.2 per cent) settlements belongs to the smallest category–A covering an area about 2000 square metre supposed to accommodating around 40 persons according to adopted norm of 20 persons per 1000 square metre. A large number of 30 (61.2 per cent) settlements enjoys a higher rank-size category–B with an approximate covered area of 5000 square metre promising shelter to about 100 persons along with their other paraphernalia. And only 3 (6.1 per cent) settlements could reach to category–C consisting of roughly 200 persons spreading over an area of about 10000 square metre. In this phase no other settlement could rise to the moderate level of category–D let alone other higher groups. It seems that the growing economic network of the mature Harappan times experienced a setback in this phase only to discourage them to establish large settlements and rush towards city centres for these were no more the harbingers of the yester-years. The number of late Harappan settlements increased in accordance with the rank-size rule according to which the number of settlements goes up as the size of settlements decrease.[22] The aggregation process began in mature Harappan times is reversed in the sense that the D and F categories represented in mature Harappan times are blank in this phase. But the substantial number of settlements representing category-B suggests that the reverse order was arrested at this level from going further down.The higher number of late Harappan sites and their association with category-B surely indicate considerable increase in the population. This increased population needed more food grains warranting opening up more pristine soils and thereby increasing the radius between their settlements and fields on account of receding chances of getting suitable locales freely. Such increase would be unviable economically after a limit even with the help of new crops like millet and thus, generate a sense of mutual antagonism among individuals for each others' possession ultimately forcing people branching off from the parent settlements to make new ones afield hitherto untouched areas.[23] This is seen in the as many as 24 fresh settlements in this phase. Consequently the resource zone of every settlement was reduced along with size of population at respective settlements eclipsing their level of organization. This reduced the operative potential of the forces of production leading to the reversal of the agglomeration process set in motion in the preceding mature Harappan levels. As a result a large number of settlements could not, however, exceed the size of small settlements like Mitathal–IIB[24], Mirzapur[25], Daulatpur[26] and Bhagwanpura[27] in the adjoining areas. And only a nominal number of 3 sites could reach the level of a moderate size village.

TABLE 4.14

PGW-GW Sites : Rank-size Frequency*

Rank Size	*Area of Sites sq. m.*	*Number of Single or Major Culture Sites*	*Population Size 20 Per 1000 sq. m.*
A	Approximate 2000	02 (3.38%)	40
B	5000	15 (25.41%)	100
C	10000	05 (8.47%)	200
D	15000	04 (6.77%)	300
E	20000	01 (1.88%)	400
F	30000	—	600
G	Above 30000	—	600

* See Appendix-II for single or major culture sites.

Rank-size Frequency of PGW-GW Sites

The rank-size frequency of PGW-GW sites in table 4.14 reveals that 27 sites fall in single or major culture representative group. And the remaining 33 sites fall outside the purview of the accepted norms for this purpose. Of them 2 (3.38 per cent) belong to the smallest category–A, 15 (25.41 per cent) to B, a category representing the major chunk of the late Harappan sites, 5 (8.47 per cent) to C, which conforms to a moderate size village. In this phase once again like the mature Harappan phase 4 settlements (6.77 per cent) reach category–D covering still large area of around 15000 square metre providing accommodation to about 300 persons unlike the late Harappan phase. A lone (1.88 per cent) site joins Category–E like mature Harappan phase. This obviously demonstrates a trend of aggregation beginning once again in these levels in so far as the rank-size hierarchy of settlements is concerned. If we compare this rank-size frequency of respective settlements with that of the mature Harappan phase the emerging picture suggests apparently a stronger tendency of aggregation in the former than the latter, which, however, remained confined mainly to the moderate size village settlements of category–C. And if the comparison is made in respect of this category the latter stands out clearly. And it runs supreme in connection with higher sized village settlements because the PGW-GW phase is better placed as the number of such settlements is a little higher than the mature Harappan phase. This shows that the tendency to gain size was stronger than the late Harappan, but weaker than the mature Harappan phase.

Rank-size Frequency of PGW-GW Sites in the Bhangar Subzones

The table 4.15 explains the rank-size distribution of PGW-GW sites in the sub-zones of the Bhangar belt. The western subzone of the Bhangar presents 9 sites of the 23 as representative settlements of this phase. These belong to different categories from A to D with 1 site to A, 4 to B, 2 to C and 2 to D. This shows that more than half of the sites in this section are covered by category A and B representing smaller villages. This makes amply clear that the rank-size frequency of settlements was dominated by the group of small sized settlements primarily.

There was, of course, a tendency of aggregation as reflected by 2 settlements each to category C and D. But this assumes the status of a moderate stimulation on the one hand and on the other provides an index of a process of gradual development of settlement hierarchy. However, this upward trend could not be sustained beyond the limit of category D.

TABLE 4.15

PGW-GW Sites : Rank-size Frequency in the Bhangar Sub-zones*

Rank	*Size*	*Western Bhangar*	*Eastern Bhangar*	*Western Margins*
A	Approximate 40	01 (11.11%)	02 (22.22%)	—
B	100	04 (44.44%)	05 (55.55%)	02 (40%)
C	200	02 (22.22%)	01 (11.11%)	01 (20%)
D	300	02 (22.22%)	01 (11.11%)	01 (20%)
E	400	—	—	01 (20%)
F	600	—	—	—
G	Above 600	—	—	—
Total		9	9	5

*See Appendix-II for single or major culture sites.

Coming to the eastern subzone we record 9 of the 22 representative sites. Of them five sites belong to category B consisting of small sized village settlements signifying their overwhelming position among the available rank-size categories. Further one site each to category C and D marks the ongoing feeble ascending order in unambiguous terms, which was somehow arrested reaching the level of category D like the previous sub-zone of western Bhangar. Besides two sites represent category A and none of the sites could attain the size of any other higher categories from E to G. And if we compare the relative number and proportion of the representative sites in the two sub-zones the eastern subzone is found densely populated like the western half. This analysis further brings out the fact that in the western subzone moderate size villages predominates, whereas in the eastern subzone small size villages of category B are more in number. Similarly category C, the most popular in the western subzone is nominally represented and so is category D in the section. This marks the difference in development processes of the two subzones.

As we move further east to embrace the Western Margins area or the eastern most extension of the Bhangar we notice five sites all along the dividing line of the Bhangar and the Khadar overlooking the Yamuna flood plain in the east. Their category-wise allocation is 2 (40 per cent) to B and one (20 per cent) each to category C, D and E. This brings out clearly the high frequency of small village settlements as usual. Interestingly this also reveals ever higher ascending order in relation to size status of available representative settlements. And consequently one (20 per cent) of them could reach the category–E like mature Harappan phase. Settlements of such a dimension could not have been established in the two other previous subzones. This development was of vital importance in this subsection, which further explains the gradual expansion of the area of human circulation from the west to the east and the advancement caused by better harmony among the operative variables. However the overall greater frequency of PGW-GW sites in the Bhangar supports Dutt's observation.[28]

TABLE 4.16

PGW-GW Sites: Rank-size Frequency in Physiographic Zones*

Rank Size		*Bhangar*	*Western Margins*	*Khadar*
A Approximate	40	03 (16.66%)	—	01 (25%)
B	100	09 (50%)	02 (40%)	01 (25%)
C	200	03 (16.66%)	01 (20%)	01 (25%)
D	300	03 (16.66%)	01 (20%)	01 (25%)
E	400	—	01 (20%)	—
F	600	—	—	—
G	Above 600	—	—	—
Total		18	5	4

* See Appendix-II for single or major culture sites.

The table 4.16 shows the division of sites into three main zones like the Bhangar, the Khadar and the Western Margins of the Yamuna.[29] The Bhangar and the Khadar are two major geographical zones, whereas the last one is a narrow dividing strip between the two contrasting zones. This categorization of representative sites on these lines speaks of greater frequency of settlements in the Bhangar belt. Most of the sites are of small size as category A and B shares 3 (16.66 per cent) and 9 (50 per cent) respectively. The former is of transient nature while others of category–B appear to have been small sedentary villages each consisting of a few families. Besides 3 (16.66 per cent) sites belong to category–C consisting of moderate size village settlements. The difference between the number of sites of category B (50 per cent) and C (16.66 per cent) is considerable given the overall numerical strength. This points to a very feeble overall current of agglomeration. The table further identified 3 (16.66 per cent) sites with category–D in order of hierarchy signifying further continuation of the enlargement process. But this seems to have terminated here only as no other settlement could assume added proportions to reach higher categories of sites. This tendency is commensurate with the then prevailling operative potential of the forces of production. Their growth beyond this level was circumscribed by such inherent constraints. Further what we witness is that along the Western Margins 5 representative sites are available. And none of them is of transient nature. The number of small sites representing category B is also only 2 (40 per cent), though it is just the double of the percentage of any other single category. And even in comparison to the combined strength of the other three categories its individual share is substantial. The remaining 3 sites, one each (20 per cent) belong to category C, D and E are indicative of a strong aggregation tendency keeping in view the overall small numerical strength of sites in a small area like this. The largest settlement produced ever by the district in this period also belongs to this small zone representing category E unlike the vast zones of the Bhangar and the Khadar. This is despite the fact that a majority of the sites are located especially in the Bhangar, a centre of human activities so far and the Khadar zone so rich in fresh water and fertile alluvial soil also possessed a good number of settlements. This leads us to enquire as to why these major zones failed to produce any settlement of the size of the Gumad representing category–E. Apparently there is no direct evidence to explain this phenomenon. However, a proper analysis of the interplay between the forces of production would perhaps unfold the full range of circumstances leading to this development. And for that matter we know very well the existing contrast between the two major zones on the one hand and the contrast between the major zones and the Western Margins. And we also know the level of operative potential possessed by the social forces. Its own inherent constraints are also not unknown. Hence, the combine of the natural and social forces was not of the order so as to accelerate the pace of development to the extent of promoting a full range of settlement hierarchy in the Bhangar as well as in the Khadar. The predominance of small and medium range settlements in the Bhangar explains its being more convenient than the Khadar, the latter being still lesser compatible with the extant operative potential, but certainly was trailing behind the Western Margins as it being a combine of the two zones proved more congenial to the available operative potential. And this was fully appreciated by the people if the settlement hierarchy is any yardstick: Keeping in view the total scenario it is difficult to agree with Lal's observation that agricultural communities in PGW levels tended to settle on fertile lands close to rivers.[30] But if this observation is viewed in the context of settlements along the Western Margins alone one would agree partially. Regarding the dissemination pattern of the PGW-GW sites in the Bhangar, Dutt underlined the dispersed nature induced by the natural drainage lines.[31] However, on the one hand there does not exist any potential regional drainage network and on the other most of the settlements are located away from such drainage lines. We have already pointed out that these settlements are available along or near some local depressions, well away from any perennial river or regional water-course of seasonal nature. And under such conditions people usually had to establish a harmonious relationship with their micro-environment in accordance with the available operative potential. Thus, a harmonious combination of these factors seems to have worked as the prime stimulant.

TABLE 4.17

PGW-GW Sites: Rank-size Frequency in Physiographic Zones*

Rank Size		*Bhangar*	*Khadar*	*Total*
A Approximate	40	3 (11.11%)	1 (3.70%)	4
B	100	11 (40.74%)	1 (3.70%)	12
C	200	4 (14.81%)	1 (3.70%)	5
D	300	4 (14.81%)	1 (3.70%)	5
E	400	1 (3.70%)	—	1
F	600	—	—	—
G	Above 600	—	—	—
Total		23	4	27

*See Appendix-II for single or major culture sites.

PGW-GW Sites: Rank-size Frequency in Physiographic Zones

The table 4.17 informs us about the general pattern of spread of PGW-GW settlements and their rank-size frequency as well in the two main geographical zones. A sizeable number of single or representative sites is concentrated in the upland Bhangar. The rank-size hierarchy displays a fair spread over categories A to E. However, smaller categories like A and B claims 14 (51.85 per cent) of the 27 selected settlements for this analysis, which points to fostering of smaller settements. A limited number of 5 (18.51 per cent) sites each under category C and D, located at strategic points in the Bhangar, tempts one to argue for a restricted stimulation to put on size. This is further vindicated by a sole (3.70 per cent) site belonging to category E, as this forms the highest class like that of the mature Harappan phase. In the Khadar belt only four sites are available for this analysis and these exhibit a consistent, yet, limited stimuli to grow in size from category A to D, though this zone is rich in natural resources and could have produced not only considerably large number, but also big sites had the natural forces alone been capable to accomplish this objective.

RW-NBPW Sites: Rank-size Frequency

The table 4.18 is dealing with rank-size frequency of RW-NBPW sites in district Sonipat. We have selected 28 sites in accordance with the accepted norm. This reveals complete absence of smaller settlements belonging to categories A to D, which means smaller settlements were not very familiar and most of the sites belong to higher categories. The highest category E of the mature Harappan and PGW levels forms here not only the smallest category, but also very meagrely represented by only 2 (7.14 per cent) sites, the next higher category F here contains 10 (35.7 per cent) sites. And for the first time the highest category of our classification is joined by 16 (57.14 per cent) sites promising accommodating to over 600 persons. This once again reinforces the previous observation that with the growing information about the properties of natural forces, technological know-how and the level of human organization and application of new crops lead to this stage of development. The size, thickness of cultural deposition and frequency of cultural artifacts altogether indicate greater economic activities than ever before. Thus, the evolutionary process reached this level traversing through various uneven stages of development.

TABLE 4.18

RW-NBPW: Rank-size Frequency in District Sonipat*

Rank Size	*Area of Site (in sq. m.)*	*Number of single or Major Culture Sites*	*Population size 20 Per 1000 sq. m.*
A	Approximate 2000	—	40
B	5000	—	100
C	10000	—	200
D	15000	—	300
E	20000	02 (7.14%)	400
F	30000	10 (35.71%)	600
G	Above 30000	16 (57.14%)	600

* See Appendix-II for single or major culture sites.

RW-NBPW Sites: Rank-size Frequency in Physiographic Zones

A comparative study of RW-NBPW sites in two different zones—the Bhanagar and the Khadar of district Sonipat has been illustrated in table 4.19.

TABLE 4.19

RW-NBPW Sites: Rank-size Frequency in Physiographic Zones*

Rank	*Size*	*Bhangar*	*Khadar*	*Total*
A	Approximate 40	—	—	—
B	100	—	—	—
C	200	—	—	—
D	300	—	—	—
E	400	02 (07.14%)	—	02
F	600	06 (21.42%)	04 (14.28%)	10
G	Above 600	05 (17.85%)	11 (39.28%)	16
Total		13	15	28

* See Appendix-II for single or major culture sites.

This indicates that most of the sites are relatively big in size as 2 (7.14 per cent) of these sites fall in category-E, 6 (21.42 per cent) in F, 5 (17.85 per cent) in G, in the Bhangar. While 4 (14.28 per cent) sites represent categories F and G is represented by 11 (39.28 per cent) sites in the Khadar. This reveals in clear terms the growing difference in the two zones. In the Bhangar we witness a fair proportional frequency of higher category sites and a gradually emerging hierarchy of settlements representing categories E to G, whereas in the Khadar an increasing tendency to prefer only large size settlements is marked. The degree and pace of growth appears to have been higher in this section than the Bhangar as the site distribution represents presentation of the two highest categories. This seems to have been induced by a higher level of symbiotic relationship among the operative variables.

Demographic Status

The aforesaid data may help us in drawing certain inferences about the demographic trends in different cultural phases. The degree of accuracy of such attempts depends upon careful survey of the size of sites, niceties in fixing their time span and conducting horizontal excavations at representative sites, which may enable us to know the actual size and internal structure of settlements.

This can further be enriched by comparing with the contemporary demographic patterns in the vicinity, the other demographic observations of past societies and their respective possible environments. In absence of horizontal excavations in our area, as elsewhere, it is extremely difficult to ascertain precisely the actual area of circulation of respective cultures by way of mere surface investigations, especially at multiculture settlements. In case of single culture sites also there are problems of their small number and disturbed state. Therefore, it is also not safe to rely heavily on the size of these sites to discern the past patterns of population growth. However, to reduce further the possibility of error the degree of sedentariness of respective settlements could be helpful[32] as all the sites representing a particular culture would not necessarily have been contemporaneous in nature for shifting more frequenctly than we visualize here.[33] In such a complex situation there exists a strong possibility of some error creeping in the delineation of size of sites, length of occupations, size and number of families and accommodation occupied. Such deformities may seriously affect population estimates and their bonafides. In the light of the above mentioned constraints our assessment regarding the past strength of population would tend to be highly provisional. Nevertheless a discussion on the available demographic models from the adjoining areas is desirable to assess their validity in the light of the settlement data from our known area and to develop, if possible an alternative model to estimate population in different periods in the area under review. In this regard one demographic model is forwarded by Frankfort. He recorded 20 houses in an area of one acre and assumed 6-10 persons per house estimating 297-494 persons per hectare with 20-34 square metre area per person. According to this estimate 24-40 persons are assigned to one thousand square metre.[34] On the other hand Fairservis based his calculations on comparative study of house plans of the past and the present village houses and estimated 6 persons per house. He assigned about 75 square metre per person and thereby 450 square metre per house. And thus, he calculated about 22 houses in one hectare with 133 persons and 13 persons per thousand square metre.[35] Adam analyzed it differently in connection with different places. According to one estimate 100 persons are given to one hectare if the size of the settlement is measured by length and breadth. And if the total area is taken into account he assigns 150 persons per hectare. On the basis of these calculations about 9 and 12 persons per 1000 square metre are worked out respectively.[36] In his second approach he takes into account the non-residential area like lanes, gardens, wells, public buildings, etc. and assigns 200 persons per hectare and 3.5 persons per family. Here about 17 persons are allotted to per thousand square metre.[37] Lal calculates 55 houses with 4.9 persons per house in one hectare for protohistoric and early historic Ganga-Yamuna Doab. In one hectare 275 persons are estimated, which bears parallel in population to those of modern villages and towns to the same size. On this basis the estimates of population per thousand square metre comes to about 23 persons per 1000 square metre and 36 square metre area per person.[38] All the above mentioned demographic models are based on tentative past and present conditions in their respective areas and their varied estimates are highly provisional. In this connection all of them appear inclined to work out a convenient working model to ensure minimum possible error in the provisional population estimates. Their calculations range between 12 to 40 persons per thousand square metre. In our own area Dutt estimated 18 persons per thousand square metre and 200 persons per hectare.[39] Keeping in view the wide range of their convenient provisional estimates and the complex conditions in our own area we propose to assume a functional arrangement to compute near possible population strength at respective representative cultural settlements. According to this working model we propose to calculate about 20 persons per thousand square metre by taking into consideration all possible factors like lanes, wells, courtyards, space for storing grains, husks, fuel items, to provide shelter to animals and room for common places. All these requirements and their provisions do constitute an integral part of village life in this area, even these days as well. The point, which needs special emphasis is the practice of animal husbandry. This has been even today, one of the essential subsistence activities of villagers. Usually people tend to maintain a fairly high number of animals for milk, ploughing and other purposes. Since it served as one of the dominant supplementary subsistence source and a predominant source of labour in absence of any effective technical know-how there becomes a greater possibility of practising animal husbandry on fairly large scale. And thereby a corresponding provision for their accommodation and fodder storage is called for. Hence, our tentative assessment of 20 persons per thousand square metre does not appear unrealistic. And according to this analysis the emergent size of population in different phases of the development is explained in the following lines. In the first phase of settlement history the western subzone offered subsistence opportunities to early Harappan people at only 12 sites. These settlements are very small in size representing the lowest category A in the hierarchy. Hence, their population size at all the individual sites does not appear more than 40 persons in any case and thereby the overall population strength of over four hundred persons is calculated who managed their subsistence from their surrounding catchment areas. However, this is not to suggest that the resource zone available between settlements and the entire western subzone was capable only to support this much strength of the people. Since this happens to be a very vital point, would be elaborated later. In the second phase, i.e., the succeeding mature Harappan phase, a minor upward trend is obvious in the number of (14) settlements representing a good range of settlement hierarchy from category A to E supporting over two thousand persons in the very western subzone. This improvement in terms of settlement number and population size is indicative of the fact that their resource zone and subsistence strategies were further enlarged during this phase. Breaking new grounds for and put on size at respective settlements strengthen this observation further. And it must be noted that there was hardly any substantial change in the quality of the resource base inherited from the preceding early Harappan phase. Another point to be taken note of is that the observed increase in population is not commensurate with the size and number of early Harappan settlements. Since the early and mature Harappan sites in the north and south of this district were not in great number the major share of the immigrants is likely to be from the western side where such settlements were already in good number (table 4.21). Besides what is more significant in this context is arrangement of food for the increased population from the same western subzone. This relates with the changing carrying capacity. And this has to be explained whether it was carrying capacity of the natural environment itself or something else, which in combination with the former constituted that. However, in the meanwhile we need to discuss the position of population in the third phase, late Harappan times.

We have already recorded the increase in the overall numerical strength of the late Harappan sites and in the western subzone of the Bhangar, which rises to 37 from the 14 sites in the mature Harappan phase. Although their hierarchy remained confined to medium to moderate size villages, yet, their collective population strength crossed the four thousand mark for the first time. This brings us to a point where the questions of the environment natural or otherwise, its carrying capacity and the interplay of the natural and social forces is more sharply focussed. Because this being the decadent phase of the mature Harappan is generally considered deficient in technological sphere and natural conditions but the higher number of settlements and population do not seem in agreement with this observation. How is that with poor technical expertize and adverse natural conditions, the late Harappan people managed to survive almost double the strength of the mature Harappan phase and that too in moderate size village settlements? Here the very premise of drastic adverse changes in ecological conditions and their overwhelming role in shaping social activities is over played. Similarly the deterioration in technical fields also needs to be examined afresh. In this context we have to differentiate between technology for subsistence and service purpose. The former kept on growing gradually and led to the beginning of production of service goods. With the disintegration of the Harappan Urban system it was this aspect of technology which suffered considerable reversals. Whereas the subsistence technology, mainly being confined to the rural settlements neither witnessed any extraordinary upsurge nor suffered any major debacle due to serious set-backs at the major urban centres, though minor tremors in the realm of subsistence technology are not ruled out altogether. However, such minor loses were partly compensated by the reversion of urban artisans along with their expertize in copper technology to the rural settlements, which is substantially displayed by an abrupt increase both in rural settlements and population. Perhaps the occurrence of Copper-Hoard objects either in the form of scattered instances or in hoards in the hinterlands is indicative of that process. Besides the very addition in population promise more hands and greater organization to further bridge up the gap created by the reversals in tool technology. And to further strengthen their operative potential they not only grouped themselves, but also devised certain new subsistence strategies. Crops like millet and sorghum said to have been introduced in this very phase fall in this category. This was further supplemented by large-scale animal husbandry and food gathering practices. Thus, their operative potential replenished in the aforesaid manner helped them to survive in the same natural conditions in large number of rural settlements.

During the late Harappan period itself we have also recorded 22 other settlements, but in the eastern subzone of Bhangar. This shows that the influence of population was tremendous to be withstood by the western subzone of the Bhangar itself. With the result a fairly large number of people, over three thousand, moved into the eastern subzone and made their habitations. However, in this context other things being more or less the same the evergrowing knowledge about the would be potential of the extant natural resources and their pristine nature played a significant role to eke out subsistence. This is what is reflected in the respective number of PGW-GW settlements in the western (23) and eastern (22) subzones of the Bhangar (table 4.15) as well. In this phase in both the zones uniform conditions seem to have prevailed to produce equal number of settlements. This situation is developed for the first time in the settlement history of this zone. In the succeeding phase of RW-NBPW the respective sites of the western and eastern subzones are 20 and 31 (table 4.2). This difference between the two subzones distrubed the balance established during the preceding PGW-GW phase. Apparently it seems as if the magnetic charm of the western subzone began somehow withering away and the eastern subzone was becoming more attractive because of some natural changes. However, our analysis of this phenomenon offers a contrasting perspective. According to which the all time high populations of about 10000 spread over in 20 bigger village settlements managed to sustain, but this was obviously neither simple nor spontaneous, rather it was the result of concerted efforts with the help of certain additions in the iron technology, subsistence practices, manpower and human organization. Moreover the eastern subzone being a little bigger than the western subzone, with shorter duration of human history and the aforesaid additions in the operative potential offered better opportunities to more people. Thus, the cumulative influence of these forces paved way for such a large number settlements in the eastern subzone.

In the next zone of the Western Margins the beginning of human history takes place with only 4 sites of small size and number of residents. No doubt the combination of the two physiographic constrating zones enriched the potential of natural resources, but the other constituents of the operative potential were not as composed as were in the western parts for being the initial stage of human history in this section. But in the succeeding PGW-GW phase the higher number (10) and size of settlements speak of greater size of population. This trend was further strengthened in the RW-NBPW levels with 18 sites with still bigger size and population. Obviously the better growth graph, higher than the western subzones is enforced by the better combine of the forces of production. However, in the Yamuna Khadar the good looking natural conditions were not of any use in the early stages of human history, from early to late Harappan, because the other constituents of the operative potential were not enough to negotiate with them, even after the introduction of iron technology in the PGW-GW levels. Only RW-NBPW levels when their operative potential was further strengthened, though still limited in comparison to the western zones, 19 settlements could be made accommodating above ten thousand people where now over 111 villages with lakhs of people are inhabited. This shows that the Khadar with fresh water, alluvial fertile soil and a range of other natural resources was full of risks than any comfort in the absence of desirable operative potential. And hence the moment the operative potential increased as in the medieval[40] and modern times the problematic looking zone became conducive to attract more and more people settle over there. This calls for their spacing and carrying capacity analysis.

Spacing Analysis

Spacing between settlements offers an interesting index of human interactions with their surroundings. And to understand the dimensions of these relations distance between settlements is calculated as crowflies. The emerging spacing patterns during various cultural phases are discussed below in order to make out the probable forces promoting spacing and thereby governing settlement patterns. However, this analysis primarily would depend upon settlement data in the absence of any other direct evidence.

Spacing in Physiographic Subzones

The table 4.20 gives us details of the existing spacing between two cultural settlements during the period under review in the physiographic subzones of the district and in different cultural phases as well. According to the emergent average distance in the western subzone of the Bhangar is ten kilometre in early and mature Harappan levels despite a higher number in case of the latter, however, this addition in the numerical strength was neutralised by a corresponding extension in their interaction zone as still more areas were available for such expansion. But the substantial increase in settlements (37) in the late Harappan levels disturbed the above mechanism to reduce the spacing level to 9 kilometre. And the process was reversed to the original position of 10 kilometre with the decrease in PGW-GW settlements to 23. In the RW-NBPW the number of settlements is further reduced to 20 only, but interestingly instead of any corresponding increase in spacing there is rather a decrease in it to 9 kilometre once again, a more interesting situation to take note of. Coming to the medieval times the average distance is reduced to 6 kilometre with the increase in settlements number to 30 out of 145.[41] And in modern times it is 4 kilometre with the increase in population and the number of settlements to 52.

TABLE 4.20

Spacing (in Km.) in Physiographic Sub-zones

Cultural Period	*Western Bhangar*	*Eastern Bhangar*	*Western Margins*	*Khadar*	*Sonipat District*
EH	10	14	—	—	10
H	10	—	—	—	10
LH	09	12	08	—	9.6
PGW-GW	10	12	08	12	10.5
RW-NBPW	09	09	07	10	8.7
MEDIEVAL	06	03	03	03	3.7
MODERN	04	02	1.5	1.25	2.18

In the eastern subzone of the Bhangar the distance during the late Harappan with 19 and the PGW-GW with 22 sites is 12 kilometre irrespective of a minor increase in the numerical strength of the latter. But during the RW-NBPW levels there is a substantial increase taking the number of settlements to 31, which naturally reduces the spacing between settlements to 9 kilometre. A similar leap in the number of settlements to 51[42] and 150 in medieval and modern times respectively further reduces the distance between settlements to 3 and 2 kilometre respectively. In the context of the Western Margins subdivision the spacing remains 8 kilometre in the late Harappan and PGW-GW levels despite their respective number being 4 and 10. Once again in the RW-NBPW levels with the further increase in the number of settlements to 18 and corresponding growth in their size the spacing pattern experiences reduction in the distance from 9 to 7 kilometre. And this downward trend is further accelerated in medieval and modern times bringing it down to 3 and 1.5 kilometre respectively with an obvious increase in their respective number of settlements. Coming to the Khadar zone the spacing is 12 and 10 kilometre in respect of the PGW-GW and RW-NBPW levels with their respective share of 8 and 19 settlements. This is further reduced to mere 1.25 kilometre in modern times with 111 villages.

Besides a look on the inter-regional spacing patterns in respective cultural phases suggests the following. Such as attempt is possible only with the beginning of the late Harappan phase mainly because in the earlier two cultural phases of early and mature Harappan the area of influence was limited to the western subzone of the Bhangar and in the late Harappan times they extended their zones upto the Western Margins, in other words enveloping the entire Bhangar zone. And during this phase the distance in the western subzone is 9, in the eastern subzone 12 and in the Western Margins 8 kilometre. This brings out an upward and downward trend from the western to the eastern subzone and to the Western Margins respectively. And almost a same pattern is witnessed in regard to the PGW-GW phase with 10 in the western subzone, 12 in the eastern subzone and 8 kilometre in the Western Margins. In the RW-NBPW phase it is evened with 9 kilometre in the first two subzones and went down to 7 kilometre in regard to the Western Margins. It is 6 in the western subzone, 3 each in the eastern subzone, the Western

Margins and the Khadar. In modern times this experiences further reduction as shown by 4 in the western subzone, 2 in the eastern, 1.5 in the Western Margins and 1.25 kilometre in the Khadar. This full range of inter-cultural and inter-zonal spacing underlines certain very important points. One of them is that in the early stages of settlement history in different zones the distance is likely to be high simply because chances of getting more pristine areas for extensive subsistence practices are available as has been observed in the context of early phase in the western subzone of the Bhangar, in late Harappan phase in the eastern subzone of the Bhangar and Western Margins and in PGW-GW levels in the Khadar. Secondly, even if there is a minor increase in the number of settlements in the second stage of settlement history the distance does not get affected partly because shifting nature of settlements and partly because availability of new areas around as has been experienced in the case of mature Harappan and PGW-GW in the western and eastern subzones and the Western Margins respectively. Thirdly, the inter-cultural analysis of late Harappan settlements in the western subsection brings out clearly fragmentation, branching off and shifting tendencies and thereby the spacing tended to narrow down substantially. Fourthly, the decrease in the number of settlements and aggregation in their size help to reduce distance between settlements in the PGW-GW and RW-NBPW levels in the western subsection. Fifthly, even with substantial addition in the number of settlements fails to produce any corresponding reduction in the spacing because the prospective downward tendency is frustrated by the process of agglomeration at individual settlements in RW-NBPW sites in the western subzone itself. And sixthly, in the medieval and modern times there is a further reduction in the spacing owing to the increase in population, number of settlements and decreasing possibilities of getting pristine resource pockets any more. In this situation constant growth of population in recent years brought the spacing to the lowest level in the Khadar. Thus, a mere look on the emerging space graph reveals minor to major variations, which apparently seem to have been induced either by the increase in population and number of settlements or by the possibility of getting additional suitable pockets around their parent settlements. The addition in the former is, irrespective of its internal or external character, part of a larger process. Similarly the very recurrent availability of additional pristine pockets within a given geographical zone does not warrant creation of new resource pockets as and when required, it rather points to the same process, which creates need for such fresh locales with the additions in the extant population, make people recognize the potential of such existing zones and enable them to harness that for their subsistence, which, however, was neither required nor feasible all the times. The growth of population and settlements, in number as well as in size, over the period, specially in RW-NBPW, medieval and modern times in the same area covered by the district under review does not reflect any overwhelming role of the natural forces in shaping the spacing pattern.[43] What we witness is a combined impact of the forces of production such as the geographical and ecological conditions, human population and their organizational level, social objectives and requirements, awareness about behaviour and potential of natural resources and the tools, techniques and strategies. Further the pace of development is subject to the mutual relations of the said variables. These are known as the operative variables and their combined impact is as operative potential. Hence, more the harmony among the variables the better is the pace of growth. However, this further needs to be examined on two other counts, the near-neighbour relations and the carrying capacity of a natural pocket in the following lines.

Near-Neighbour Relations

The foregoing spacing analysis is useful to understand the past near-neighbour relations among the residents. These relations seem to evolve on two levels—continued traditional bonds and the circumstantial emerging new contacts. The former represents their traditional roots and the branching off process from the parent settlements to make fresh settlements in close vicinity for a variety of factors to be addressed to later. This in fact is an integral part of a bigger process, the gradual advance movement of the people. And their dispersal in a contiguous and proximal manner form hallmark of their mutual relations. According to which, since the inhabitants of the new settlements have splitted up from their parent settlements they tend to retain that identity and association alive for years together and this continues to reinforce their traditional bonds. This is the primary level. Whereas the latter involves their mutual new circumstantial relations. Such relations are generated by proximity of their dwellings and fields,[44] transhumance, especially in summer seasons, village fairs, tournaments, feasts on different occasions,[45] etc. Besides the practice of extensive grazing year in and year out, so essential for largescale stock raising, collection of wild fruits, roots, gum, honey, green fodder, fuel, game, clay, etc., from specific locales and in different seasons also necessitate to foster mutual contacts. This is to gain intimacy, to get acquainted and benefited with each others possessions and to develop a sense of association and antagonism. In addition in the face of hard natural conditions and technological constraints people knew their continued limitations about their operation potential which created awareness about the relevance of cooperation and coordination among them. This resulted into a considerate and reciprocal attitude among the people in general and residents of closeby settlements in particular. As a result people of the nearby villages seem to have shared and lend food grains, tools and implements, draught and milk animals, fodder and seeds. They also assisted in terms of manpower, if not hire labour, from time to time as the situation arose. There could have been several other such levels as well to enlist the support of more contemporary people to safeguard against natural and human threats. Currency of such practices among the contemporary societies in and around this area could lend further credence to the validity of the present observations in such a situation. Therefore, the distances between settlements do not seem to pose any insurmountable problem for the people in ancient times to interact among themselves, especially in absence of any better communication means, as was the case in recent past.

Carrying Capacity and Splitting up Tendency

The concept of carrying capacity denotes an advance stage in the study of settlement patterns. This has been interpreted differently by people. According to one observation once the population density reaches the maximum level of the carrying capacity of land it cannot be increased by mere extra labour input unless any innovation in the realms of technology or occupation of pristine land is effected.[46] This suggests a strong need either to improve the existing technical know-how or to occupy a fresh piece of land to increase food production in order to support increased population density. Here the role of technology is rightly underlined, but it surely underestimates the productive potential of additional human resource and also further confuses the carrying capacity of a piece of land/pocket/zone as a unit by suggesting occupation of pristine land to improve production because the very concept of carrying capacity is strictly linked to and measured of a fixed area. While the other observation emphasises the exploiting capacity of a group on the one hand and links it to the extension of the perimeter of the catchment area. And further underlines the need of harmony between them to keep carrying capacity promising. As soon as the harmony is disturbed with the perimeter of the catchment area going beyond the economic range of their exploiting potential both the carrying capacity of the catchment area and the exploiting potential would tend to deteriorate rather than improve.[47] In this context a correlation is sought to be established between the better operative potential and the extending catchment zone to assess their carrying capacity. This relationship, however, does not stand scrutiny because the ineffectiveness of the extant technology and non-availability of regular suitable geographical locales, among other things, promote extensive farming and thereby enlargement of the catchment zone is against the very spirit of carrying capacity of a locale. Therefore, keeping in view these observations and the environment of our area it becomes imperative on our part to understand first the import of these concepts of catchment area and its carrying capacity in the following paragraphs.

The concept of catchment zone is very misleading and may vary from area to area, time to time and society to society as at times it envelopes far-flung areas providing essential goods and services. In this situation one finds difficult to establish any meaningful relation between the catchment zone and its carrying capacity. However, in relation to the present exercise in our area we take the spacing between two settlements as the immediate primary catchment zone. This zone presents a natural resource base consisting of diverse soil types, bio-diversity and ecological forces. All these constitute a 'natural environment', which conforms the resource base of the catchment zone. Here the term 'natural environment' is used deliberately to denote a specific context and meaning. This means what we receive in original form from the nature without any manipulation by human society is the 'natural environment' and what is transformed by human efforts is an 'environment'. This distinction between the two is useful to understand the factual carrying capacity of the respective environments. Hence, the carrying capacity need not be viewed in isolation, rather in conjunction with the operative variables, which when taken together amounts to the carrying capacity in the practical sense of the term.

We have already seen above in the context of the western subzone of the Bhangar from early Harappan to modern times that the size of the subzone being constant witnessed increase in the number of residents over the period. Similarly in the other subzones also this trend is repeated. And in the Khadar zone too with the beginning of habitations the upward growth of population continued till modern times. This makes amply clear that the carrying capacity of the environment gradually increased with the growing operative potential of the people. Thus, in the same geographical zones we encounter the two extremes, the lowest and the highest levels, of population growth pattern, which do speak of the changing levels of the carrying capacity and the harmony between the natural and social variables. In this situation the degree of the carrying capacity of such an environment largely depends upon the level of harmony among the said variables. If the harmonious level is below normal caused by under development of the constituent variables the carrying capacity would also in all probability tend to be affected in accordance with, though the impact of minor imbalances in the constituent variables may not be quite discernible always in the archaeological data, albeit the influence of major-imbalances would be quite obvious in terms of expansion or contraction of settlements and their cultural artifacts as has been observed above. And thus, the number of persons to be fed in a given space and time with the help of the effective strength of the environment must be conceived as the carrying capacity. The absence of Harappan sites in the Khadar and fewer number of settlements in the PGW-GW and RW-NBPW levels on the one hand and higher number of settlements in medieval and modern times on the other was not because the extant features of this zone were absent in ancient times, but because the other variables of the operative potential were not potent enough to utilize the 'natural resources' offered by it in the early phases of human history. And the changes in the number of sites in subsequent phases is an index of the improving effective operative potential and the growing harmony among the forces of production. This harmony is also disturbed by the overgrowth of social forces, as in the case of modern society in this very zone itself, affecting adversely the health of the natural forces, which may lead to a stage where every improvement in the former would produce diminishing returns. Thus, the secret of higher carrying capacity lies with the optimum balance between the operative variables. And the shift in the balance on either side tend to affect the attitude of the residents to stay or march on to some suitable locales, which brings in the problem of branching off tendency as is discussed below.

In both the cases of minor and major imbalances there is a possibility of splitting up of the settlements with a view to search another suitable locale for making a new settlement so that the required level of food production is maintained. This is demonstrated by first inhabited 8 (47.05 per cent) settlements in the mature Harappan times in the western subzone of the Bhangar, 35 (58.33 per cent) late Harappan phase in both the subzones (western and eastern) of the Bhangar and the Western Margins, 31 (49.20 per cent) PGW-GW sites, 53 (60.22 per cent) RW-NBPW sites in the Bhangar and the Khadar (table 4.21). The average percentage of these newly inhabited sites fluctuates between 45 to 60, which apparently seem to have related with the change in the number of settlements. However, this must not be construed always as part of mere splitting up process stimulated by saturation of the carrying capacity with the gradual growth of population at respective settlements. Such increase is also caused by influx of population, among other things, as was the case in late Harappan times. It is also suggested that in semi-arid areas any setback to crops owing to frequent inherent fluctuations in rainfall system may not only cause numerous hardships to people, but also force them to bud off from the parent settlement, especially in absence of provisions of irrigation, storage and transportations. Such fluctuations in the rainfall pattern are integral to the climate of this area, which at times assume serious proportions too resulting into drought and famine like conditions as was in 1987-88.[48] A mere glance on the past available rainfall history would further endorse that. No doubt such conditions were fraught with many hardships to which human society has been subjected to. But even in such tough times what we witness is a remarkable resistance shown by people to hold back at the parent site, despite recurrent temptation to march on to some unknown, but supposedly better place, until they were pressed hard to the hilt. The 15 of the 23 PGW-GW sites in the western subzone point it clearly.[49] Obviously people usually did not possess, entertain and promote a temperament of geographical mobility among themselves for various reasons. To cope up with these conditions people resort to collection of wild fruits, roots, rough grains, hunting, fishing and stock raising. Besides the prevailing rainfall trends unfold a vast range of incomprehensible variations, which at microlevel are hardly visible in official figures for these are collected from widely located gauge centres. Moreover the localized nature of famines is also well-known in this area (tables 4.4-7) as shown by variations in rainfall at local and regional levels.

TABLE 4.21
Cultural Frequency at Individual Sites in Physiographic Zones

Cultural Phase	*Bhangar*	*Khadar*	*Total*
EH	3	—	3
EH+H+LH	5	—	5
EH+H+LH+PGW-GW	1	—	1
EH+H+LH+PGW-GW-BSW	1	—	1
EH+LH	2	—	2
EH+LH+PGW-GW	1	—	1
EH+RW+MED	1	—	1
H+LH	8	—	8
H+LH+PGW-GW+RW-NBPW	1	—	1
H+LH+PGW-GW-BSW+RW-NBPW	1	—	1
LH	12	—	12
LH+PGW-GW	9	—	9
LH+PGW-GW-BSW	3	—	3
LH+PGW-GW+RW-NBPW	4	—	4
LH+GW-PGW	6	—	6
LH+GW-PGW+RW-NBPW	2	—	2

Cultural Phase	*Bhangar*	*Khadar*	*Total*
PGW-GW	6	—	6
PGW-GW+RW-NBPW	2	1	3
PGW-GW-BSW+RW-NBPW	2	—	2
PGW-GW-BSW	3	—	3
PGW-GW+RW-NBPW	2	—	2
GW-PGW+RW-NBPW	6	3	9
GW-PGW-BSW+RW-NBPW	1	1	2
GW-PGW+RW-NBPW+Med	3	1	4
RW-NBPW	6	1	7
RW-NBPW+Med	32	12	44
RW-NBPW+Kushan+Med	1	—	1

On the basis of these past and present trends no uniform grave conditions can be visualized over the whole landscape forcing people to march on lock stock and barrel in search of better locales to settle down. And that is what we witness in regard to PGW-GW levels where a movement of people is outlined within this area with gradual advancement. And this was further strengthened in the succeeding RW-NBPW levels. Thus, people as it appears, apparently had accepted the climatic conditions as part of their life-style. To face successfully such natural conditions coordinated efforts were called for to realize their common objective of managing subsistence by labour and crop strategy intensive subsistence methods. And for that matter pooling down of more human resources was of utmost interest, which was, however, feasible only when people generally stay back at their parent sites. This is supported by the average small size of new sites in general suggesting some people exercised their last option to move out of the parent sites[50] for in the absence of matching resource use techniques,[51] frequent pockets of elevated aeolian loose-soils and small-scale extensive farming with inherent fallow system[52] extended the distance between the homestead and fields beyond economic range. With this increase in the radius the degree of connectivity declines shortening working hours and consequently hampering the possibility of improving the level of food production. Such a situation further lead to under utilization of extant environment, increased wastage of time and energy, frustrated initiative and ultimately proving strenuous to cover large distance from the residence to work place and back.[53] This also failed people to safeguard their crops from being damaged by stray and wild animals. Being close to one's holdings one cannot only afford easily to put in more labour, but also can devote more timely attention for day to day lookafter, harvesting and thrashing. It further helps them to assert their control over land in order to protect it from being damaged by ever active natural forces such as erratic rain and wind erosion, both very common and powerful agents of soil erosion. Appropriate arrangements to hold rain-water in the fields by bunding method help increase soil fertility as this induces to maintain higher water contents in crop root zone of soil and low soil temperature for longer periods, so crucial to raise good crops. These factors considerably enhance soil resilience to absorb shocks, quick climatic fluctuations, so common in the crop maturing months of October-November for *Kharif* and March-April for *rabi* crops. These periods incidentally strictly conform to the transition period from monsoon to winter and to summer. In this context the transition from one rainfall zone to the other is also crucial. Such an arrangement provides enough scope for smooth natural growth and maturing of *barani* (dry) crops to ensure stable yields.

However, one can legitimately argue as to how people managed to arrange water to meet their requirements in times of rain failure? Obviously it is difficult to explain it in a fairly convincing manner because there was no canal system whatsoever, no perennial river to name and above all the hydrological regime was most inhospitable for saline nature of soil water. It becomes more difficult, if one is devoid of first hand information of the existing micro-ecological network so crucial to understand processes of past societies. If one traverses through the area there does not seem any obvious permanent natural source or provision of water to fall back upon. Nevertheless the answer to this vexed problem could be sought with the help of prevailing ecological and topographical conditions on the one hand and on the other through the very locational pattern of ancient settlements. This can further be corroborated with the help of current practices of obtaining water by the contemporary inhabitants of this area. We have observed already while dealing with the locational analysis that the ancient settlements are found located without fail on or along natural depressions of varying proportions. And these formed the central hub or main lifeline of their successful existence. But how is that these depressions managed to retain water to serve requirements of the people? What was the source of supply of water to these depressions? The answer to the second question is that the supply line was mainly consisting of rain-water, since, obviously, there was no other provision. And, therefore, during the monsoon season rain-water from the surroundings used to have accumulated in these depressions. Thus, these depressions formed the repository of surplus rain-water to be used rationally so as the required quantity of water is maintained till the next rain begins. The explanation to the first question lies with the mechanism of retaining water in the ponds for most part, if not the whole year. First there is every possibility of being these depressions of greater proportion in the past than what we witness these days. Obviously the process of silting up of these depressions is continuing since ancient times by way of soil erosion caused by rain, wind, grazing and farming activities. If we presume relatively a bigger size than they possess today to store greater amount of water, then the mechanism of retaining water supply becomes obvious. This was possible by maintaining the equilibrium between the rate of loss of water by way of percolation, evaporation, transpiration, etc. and the available amout of water in these depressions and the possibility of their refreshing from time to time. To be precise we can state that as a rule these depressions contain clayey soil caused by deposition of argilaceous matters from the surrounding slopes with the flowing rain-water[54], which produces a coating of fine silt on the bottom surface plugging soil pores to a great extent. As a result such soils do not enjoy any appreciable porosity and thereby lose permeability substantially. Consequently, the loss of water through this process is drastically minimized. Moreover there was no provision for exploitation of soil water at any appreciable rate till very recent past. This rendered the soil riplet with high water content and consequently bringing the water table close to the surface, which further mitigated opportunity of any further loss of water through leaching process, because the pores of soil were already filled with water. Secondly the depressions being permanent natural water repositories were bound to support relatively a higher vegetation growth than those of the rising aeolian pockets and level loam soils. This was supposed to prohibit the adjoining land to come into direct contact with the sun, which helped in a big way to maintain low surface and undergound soil temperature. This reduced the rate of evaporation and transpiration. Thirdly, the prevailing annual rainfall patterns promise refreshing of the available amount of water at intervals by way of cyclonic year round recurrent showers of varying intensity. Thus, in normal conditions there was a strong possibility to retain water throughout the year (figs. 5.54-56).

Now we can try to understand as to how the people in the ancient past did manage the supply line in abnormal drought times? Was it really feasible to survive such natural onslaughts under the given situation? It is really a mind boggling problem and is difficult, if not impossible, to explain it satisfactorily. However, once again the existence of natural depressions with ancient settlement remains around them and the ethnographic data provided by current life-styles of people in the area constitute a perennial source of information to unravel such mysteries of the past. In periods of setbacks to prevailing rainfall system there was every likelihood of drying up of these depressions causing acute shortage of water to be managed. This problem seems to have been further compounded by the prevalence of saline subsoil water regime. Under these unfavourable conditions these very depressions seem to have provided some avenues to arrange supply of water in some form or the other. The ethnographic approach to this problem suggests sinking of some *kuccha* wells either on the banks or inside these depressions as is the common practice among the villgers in the entire area of old alluvium to which the Bhangar part of the district under discussion, the centre of activities, forms an integral part. In these depressions chances of getting fresh water exist, especially in the upper strata alone, caused by constant leaching of fresh water from the stagnant pond water into the shallow hydrological regime for year in and year out. In the process the existing salt contents from the upper layers of the soil are washed down to transform the quality of the water from saline to potable. The quality, quantity and depth of such a reserve correspond to the degree, horizontal and vertical impact of the leaching activity. The reserve of the, thus, transformed water could have been used by the people in the ancient past as was being done by the people in the recent past and is being done in modern times as well. Hence, the low-lying areas promised some amount of potable water to brave off the ongoing drought. This hypothesis is supported by the location of ancient settlements and their extension patterns along the longer axis of these depressions. Moreover in this area, especially the upland Bhangar dry belt, most of the modern settlements receive their drinking water from such wells, though now *puccka* or brick-lined, located on an average in and around village ponds (figs. 5.54-56).

In the light of the aforesaid factors one may not tempt to conceive climatic changes as the predominant factors to encourage people for desertion of their parent settlements in the Bhangar belt. The overwhelming tendency of geographical immobility is still quite prominent among residents of this area. In regard to this one may argue that these days availability of better tools and canal irrigation facilities contribute to hold them back as there is no derth of water for drinking as well as for irrigation. Such a situation on the one hand reduces gravity of droughts and on the other enhances the carrying capacity of soils, which together does not inspire people to bud off from their parent settlements to establish new ones. But contrary to it we may recall that the introduction of effective canal systems does not enjoy a fairly long history. This system began in AD 1798 only for all practical purposes, if we exclude such attempts in early medieval and medieval times by some rulers.[55] And the process of dissemination of canal network is in no way different than the slow and uneven spread of overall development programmes. And so far the technical expertize is concerned, it is also of recent origin, which received special momentum with the introduction of green revolution in the mid-sixties of this century. Before that the technological scenario was primarily not very healthy for its being consisted of traditionally simple and limited tool-techniques demanding more of human labour and time. This does not seem substantially different from that of the ancient past. Thus, people being ill-equipped like their counterpart in the past managed to survive in a more or less similar environment fraught with countless hazards while sticking at their parent sites.

Of the other factors promoting splitting up tendency among the people mention may also be made of epidemics. Such natural calamities were not unusual till the very recent past in the absence of suitable medical facilities, balanced food and civic awareness. These epidemics at time assumed enormous dimensions forcing people desert their settlement either temporarily or for ever. The former hardly leads to addition of any new settlement than the latter. While the latter provides possibilities to add the number of archaeological settlements in the form of deserted and new settlements. Only in the third possibility of partial permanent desertion there exists a scope to increase the number of sites by making new settlement. However, this cannot be construed as the prime factor to promote settlement patterns for its inherent limited impact in terms of intensity, duration and location. However, the size of new sites is in conformity with this trend.

In some cases self-assertion and competition of male adolescent with their elders for independence may inspire to branch off from the parent site and establish a new independent settlement. Such a trend is fairly visible among modern residents of the Khadar belt. But its genesis lies with the growing operative potential. Because of this change people feel quite confident to lead independent life as there does not seem any danger to their life and livelihood independent of the parent settlement. As a result despite the usual gravity of the river floods, the entire Yamuna basin including the western line of the active river course is beaming with interspersed settlements of varying size unlike the ancient past. Therefore, this development is of recent times and a similar situation cannot be visualized in the context of ancient times.

The foregoing discussion highlights that, whenever people were ill-equipped in terms of operative potential and had to face intractable natural conditions they tend to put up a very cohesive front to survive the natural onslaughts. And such conditions offered a very limited latitude for any independent initiative in asmuch as making a new settlement. On the other hand as and when people gained access to better operative potential they began harnessing the intractable looking conditions, aspiring for and asserting their independence to the extent of making new settlements. Because with this development the constant binding force–the need of close cooperation induced by the natural threats in the absence of matching operative potential evaporates quickly.

The Concept of Epicentre

The term 'nuclear zone' was first used by late Subba Rao in his book, *The Personality of India*[56]. Afterwards this attracted growing attention to remain a centre of debate in Indian Archaeology. To explain the term some rely heavily on geographical factors alone[57], while others argue that in absence of suitable technology the geographical factors alone are hardly of any use to human society.[58] However, to understand the realistic and pragmatic side of the term we need to incorporate some more essential aspects like ecology, human organizational strength and the pragmatic subsistence strategies adopted by the people over the period to meet their requirements, which in turn helped sustain a sizeable number of people in a particular zone. The intensitiy of human activities in terms of area and period would qualify any area of being an epicentre. Hence, wherever these factors are in optimum harmonious position, would in all probability tend to promote intensive production activities and human concentration turning some areas more thickly populated than the others. Such zones with intensive production activities can, therefore, be termed nuclear zones. And this is what we encounter in regard to the Bhangar in general and the western subzone of the Bhangar in particular right from the beginning of human history in early Harappan times down to PGW-GW levels. This is attributed to the combined effect of the aforesaid factors. And with the transformation of the said factors the hitherto inhospitable Khadar zone also began beaming with human activities. This signals that factors other than natural were also equally significant to promote growth of production, population and settlement patterns. Such a potential process is discernible even among the contemporary settlers in consonance with the distinctly transformed operative potential of the people. With the result the same intractable zone of the Khadar has been converted into the most thickly populated zone not only in the district, but in the state as well.

Regarding the clustering of late Harappan settlements Chakrabarti's observation of non-existence of any settlement to be a centre of cluster of smaller sites in its immediate neighbourhood[59] does not withstand in view of the recent discovery of several mature Harappan and late Harappan sites of varied denominations in close proximity with one another around Ghadwal-Rindhana complex. We, in fact, do not know about their precise nature and function at present. And, therefore, it is difficult to make out if any of these formed an enclave of agricultural villages adapted to either rural environment or constituted an administrative, productive, distributive or commercial centre. Nevertheless if the size combination, fairly large currency of semi-precious beads (figs. 5.43-44, 64-69), faience bangles of different shapes and sizes and pottery types are any indication, it may not be out of place to assign them either local marketing or distribution functions. In the light of the greater frequency of pottery at Rindhana-II, this can either be considered a centre of production of pottery or its distribution among the several settlements in close proximity. However, there is no direct evidence of mould, unfished bead and turntable except potters terracota dabbers (fig. 5.72:1-3) usually used in pot making to support our observation. Therefore, it remains provisional until or unless such other concrete evidences are obtained.

Table 4.22 displays the district wise frequency of sites in respective cultural phases in Haryana. This underlines the shifting concentrations of settlements in different districts as well as in geographical zones. During the early Harappan times the districts of Bhiwani, Rohtak, Hisar, Jind, western parts of Karnal and Sonipat formed the main area of concentration. And these very parts remained a focal point, even during the succeeding mature Harappan times, though with diminished number of settlements. Interestingly these very areas conform with the Bhangar zone. In the late Harappan phase the number of settlements goes up to 637 bringing into human fold districts like Ambala, Kurukshetra Gurgaon and central parts of Karnal and Sonipat. This was further extended to Faridabad, eastern parts of Sonipat, Karnal,

Kurukshetra and Ambala districts in PGW-GW levels. This suggests gradual eastward movement of the people and also explains the changes in their areas of concentrations. The table further brings out clearly the process of sequence and spread of habitations, which clearly underlines an eastward thrust from the western districts of Hisar, Bhiwani and Jind to Rohtak, Sonipat, Karnal, Ambala, Gurgaon and Faridabad right from early Harappan down to early historical times. This is in consonance with our observation in this regard.

TABLE 4.22

District-wise Ancient Sites-Haryana

District Name	*Culture* EH	H	LH	PGW-GW	RW-NBPW	Total
	(1)	(2)	(3)	(4)	(5)	(6)
Ambala	00	04	31	72	24	131
Bhiwani	29	06	11	04	05	55
Faridabad	00	00	01	24	07	32
Gurgaon	08	02	114	79	40	243
Hisar	29	27	36	29	02	107
Jind	45	27	131	60	18	272
Karnal	10	11	65	60	27	173
Kurukshetra	03	17	75	167	45	307
Mahindergarh	02	00	04	00	00	06
Rohtak	42	12	106	48	54	251
Sirsa	01	03	03	20	05	32
Sonipat	12	17	60	63	88	240
Total	181	126	637	626	315	1849

Conclusion

The protohistoric settlement data have been analyzed from different angles in the foregoing discussion and the emerging patterns can be summarized in the following lines. This shows that the inception of settlement history in this section is associated with the appearance of early Harappan settlements sometime in the second half of this phase. Its continuation rather consolidation is verified by the presence of relatively higher number of large-sized mature Harappan settlements in the succeeding period. The locational analysis of settlements of these two periods argues for their concentration mainly in the western section. This is again repeated in the following late Harappan phase, but surely this time with an obvious eastward dispersal of settlements in the area between Gohana in the west and Sonipat in the east, albeit more in the form of an underlying current. In the PGW-GW phase the ongoing eastward advancing process is further consolidated by bringing the settlements share of west and east Bhangar at equal level. This balance is titled in favour of the eastern Bhangar in the succeeding stage of RW-NBPW with appearance of increasing number of settlements. A similar tendency is strongly reflected in regard to beginning of settlement history in the Khadar bed of the Yamuna. Where the process of habitation as such began in fact in the later levels of the PGW-GW and seems to have been consolidated only in the RW-NBPW levels. We have already observed that this was the result of interplay between the operative variables and none of them alone seemed to have wielded the driving force, rather the process and pace of developments depended upon the level of harmony among them.

NOTES AND REFERENCES

1. Adams, R.M., *Heartland of Cities*, Chicago, 1981, p. 52.
2. Sharma, M.M., *Archaeology of Ambala and Kurukshetra Districts, Haryana*, Ph.D. Thesis (unpublished), Kurukshetra University, Kurukshetra, 1978.
3. Singh, A., *Archaeology of Karnal and Jind Districts, Haryana*, Ph.D. Thesis (unpublished) Kurukshetra University, Kurukshetra, 1981, pp. 66-67.
4. Phogat, S.R., *Archaeology of Rohtak and Hissar Districts, Haryana*, Ph. D. Thesis (unpublished), Kurukshetra University, Kurukshetra, 1978.
5. Thakran, R.C., *The Sequel to the Indus Civilization in Greater Indus Valley*, M. Phil. Dissertation, (unpublished), Jawaharlal Nehru University, New Delhi, 1981.
6. Singh, 1981.
7. Dutt, B., *PGW Settlements in Haryana*, Ph.D. Thesis, (unpublished), Kurukshetra University, Kurukshetra, 1980, p. 168.

8. Singh, 1981.
9. Dutt, 1980.
10. Adams, 1981, p. 7.
11. *Ibid.*
12. Thakran, R.C., 'Protohistoric Archaeological Remains of Union Territory of Delhi', *PIHC,* Calcutta University, Calcutta, 1991, pp. 808-16.
13. Gaur, R.C., *Excavation at Atranjikhera*, Delhi, 1983, p. 123.
14. Lal, M., 'Summary of Four Seasons Explorations in Kanpur District' *ME*, 1984, p. 64.
15. Agarwal, D.P., *The Copper Bronze Age in India*, New Delhi, 1971, p. 228.
16. Dutt, 1980, p. 181.
17. Hodges, H., *Technology in the Ancient World*, New York 1974, pp. 50-51.
18. Chitalwala, Y.M., 'Harappan and Post-Harappan Settlements Patterns in Rajkot District of Saurastra', in Agrawal, D.P. and D.K. Chakrabarti (eds.), *Essays in Indian Protohistory*, 1979, pp. 94-114.
19. Hole, F., 'Social Formations in Western Iran 8000-4000 BC,' in Binford and L.R. Binford (eds.), *New Perspective in Archaeology*, Chicago, 1968.
20. Shaffer, G.J., 'The Protohistoric Period of Eastern Punjab: A Preliminary Assessment', in Dani, A.H. (ed.), *Indus Civilization New Perspectives*, Islamabad, 1982, p. 95.
21. Bhan, S., 'Early Settlements in Haryana: A Case Study in Urbanism', *JHS,* VIII, Kurukshetra University, Kurukshetra, 1976, p. 5.
22. Brush, J.C., et. al., *Geographical Review-45*, 1955, pp. 599-619.
23. Earl, T.K., 'A Nearest-Neighbour Analysis of two Formation Settlement Systems', in Flannery, K.V. (ed.), *The Early Mesopotamian Village*, London, 1976; pp. 196-223.
24. Bhan, S., *Excavations at Mitathal (1969) and Other Explorations in the Sutlej-Yamuna Divide*, Kurukshetra University, Kurukshetra, 1975, pp. 8-9, Idem 'Transformation of Harappa Culture in Haryana', U.V. Singh (ed.), *Archaeological Congress and Seminar: 1972*, Kurukshetra University, Kurukshetra, 1976, pp. 23-30.
25. *IAR,* 1975-76, p. 18.
26. *IAR,* 1968-69, pp. 8-9.

27. Joshi, J.P., *Excavations at Bhagwanpura 1975-76,* ASI, New Delhi, 1993, p. 9.
28. Dutt, 1980, p. 168.
29. Thakran, R.C., 'Western Margins of the Yamuna and Protohistoric Cultures: A Case Study of District Sonipat of Haryana and Northern Parts of Delhi', *PIHC,* Dharwad, 1989, pp. 648-56.
30. Lal, M., *Rise of Civilization in Ganga-Yamuna Doab*, Delhi, 1984, p. 166.
31. Dutt, 1980, p. 168.
32. Smith, P.E.L., 'Land Use, Settlement Patterns and Subsistence Agriculture: A Demographic Perspective', in Uckko, P.J. et al (eds.), *Man, Settlement and Urbanism*, Duckworth, 1972, p. 424.
33. Dutt, 1980, 181.
34. Frankfort, H., 'Town-planning in Ancient Mesopotamia', *Town Planning Review* 21, 1950, p. 98-115.
35. Fairservis, W.A., 'Origin, Character and Decline of an Early Civilization', *American Museum*, Novitates, 2302, American Museum of Natural History, 1967, pp. 1-48.
36. Adams, 1981, p. 69.
37. Adams, R.M., *Land Behind Baghdad*, Chicago, 1965, pp. 23-25, 41.
38. Lal, 1984, p. 24.
39. Dutt, 1980, p. 177.
40. Thakran, R.C., 'Investigating Medieval Settlement Patterns in Sonipat District, Haryana, *PIHC*, Calcutta, 1996, pp. 972-82.
41. Jarman, 1972, pp. 61-65.
42. Adams, 1981 p. 63.
43. Lal, 1984, p. 189; Dhavalikar, M.K., 'Culture-Environment Interface: A Historical Perspective, *PIHC*, New Delhi, 1992, pp. 1-24.
44. Layton, R., 'Settlement and Community', in Uckko, P.J. et.al. (eds.), *Man, Settlement and Urbanism*, Duckworth 1972, pp. 377-81.
45. Tonnies, F., *Community and Society*, Michigan, 1957.
46. Bridsell, J.B. 'A Preliminary Report on New Research on Man-Land Relations in Aboriginal Australia', *Population Studies in Archaeology and Anthropology : A Symposium for American Archaeological Memoir*, 1982, pp. 34-37; Smith, 1972, p. 411.
47. Vita-Finzi, C. and E. Higgs, 'Prehistoric Economic in the Mount Carmel area of Palestine: Site-Catchment Analysis', *PPS,* 1970, pp. 1-36.

48. *The Times of India*, New Delhi, dated 23.4.1991.
49. Thakran R.C., 'Dynamics of PGW Settlement Patterns in the Old Alluvium of Sonipat District *PIHC*, Aligarh, 1994, p. 866.
50. Smith, 1972, p. 411.
51. Hodges, 1970, pp. 50-51.
52. Allchins, F. and B.R., *The Rise of Civilization in India and Pakistan*, Delhi, 1983, p. 141.
53. Hodder, I. 'Some New Directions in the Analysis of Archaeological Data at the Regional Scale', in Clarke, E.L. (ed.), *Spatial Archaeology*, London 1991, pp. 119-40.
54. Limbrey, Susan *Soil Science and Archaeology*, London, 1975, p. 230.
55. Joseph, E. *Third Regular Settlement Report of Rohtak 1905-10*, 1910, pp. 1-2.
56. Subbarao, B., *The Personality of India,* Baroda, 1958.
57. Adams, 1981; Lal, 1984, pp. 199-200.
58. Sharma, R.S., *Material Culture and Social Formations in Ancient India*, Delhi, 1983, p. 10.
59. Chakravarti, D.K., 'Size of Harappan Settlements', in Agrawal, D.P. and D.K. Chakrabarti (eds.), *Essays in Indian Protohistory*, Delhi 1979, p. 209.

5

MATERIAL REMAINS

In the process of intensive combing of the area under discussion for the purpose of this work a large body of protohistoric and early historical settlements, representing a sequence of cultures in succession, was put on the archaeological map of this area. These settlements put forward a vivid cultural complex, which is to be discussed in the following paragraphs with a view to understand the nature of cultural remains at various stages in the period under debate, the process of cultural developments and their significance in the life of the people at various points of times. This, in short, may lead to help explain the techno-socio-economic levels cherished by people during course of time.

Any instant survey of the gathered material would immediately reveal the composition of the material culture of this period. On the basis of these very cultural remains one would suggest that the main constituents of the material culture are pottery, terracotta objects, stone artifacts, faience articles and some other items of miscellaneous nature. Keeping in view this cultural composition and the yardstick of relatively higher frequency one would tempt to presume a predominant position for pottery among the cultural variables. This has usually been observed in context of the protohistoric cultures, especially when studies of these cultures are basically confined to surface collection in absence of excavations. In the following discussion, therefore, what is bound to follow is greater emphasis on pottery, but this would not in any way to underplay other cultural artifacts. Because these very artifacts are also bonafide components of the cultural whole. Therefore, the objective of this exercise is to bring out a near realistic picture of the cultural activities and achievements in the period under review in a sequential manner. This is to be dealt with period-wise in what follows:

Early Harappan Pottery

The early Harappan pottery have already been recorded from various excavated settlements like Kalibangan[1] in Ganganagar district of Rajasthan, Rakhigarhi, Siswal[2] and Banawali in Hisar district[3], Mitathal in Bhiwani[4], Balu in Jind district[5] of Haryana. These sites present a rich variety of shapes and painting designs of this pottery–which in turn demonstrate a gradual evolution of the art of pot making in early Harappan times. The early Harappan pottery gathered from the area under discussion is indicative of the further eastward extension of that tradition. No doubt the pottery bears close association with that of the type-sites in regard to their shapes, decoration and surface treatment. This has been amply demonstrated in the colour scheme—usually decorated in black and chocolate colours over plain, occasionally polished, red surface–thin section and shapes. But at the same time it surely lacks in variety of shapes, painting designs and surface treatment. Nevertheless keeping in view the overall family likeness this is assigned to the same polychrome group of early Harappan pottery. And on the basis of regional comparison it is very close to the Siswal group of early Harappan phase.[6] Therefore, it can be placed in terms of time period with that of early Siswal-B—the second half of the early Siswal group. A detailed survey of the available pottery is given below.

The early Harappan pottery from this area is primarily of red group. This ware is usually devoid of any slip or wash. Occasionally either red slip or red wash seems to have been applied. But that too is of uneven and fugitive nature. The pattern of its application is not uniform invariably. Moreover, it is strictly restricted to the rim, neck and shoulder portions. As a result the lower part is either plain red or rusticated. And this type of practice is not unique in any manner, since similar fashion is also betrayed at the type sites mentioned above. It is of thin to medium fabric, wheel-turned (though in certain instances the surface is uneven indicating either poor workmanship or perhaps being thrown on slow moving wheel or more likely a sign of handmade pottery), fairly well fired giving it a pinkish to red colour and made of well–levigated clay.

However, the fine preparation of clay does not seem to have been a permanent feature for the presence of calcium carbonate pieces in section of potsherds form a recurrent feature.

Regarding the decoration mechanism a synthesis of painting and incision techniques is indicated. The former is usually executed in black colour, though occasionally chocolate colour is also applied. The painting designs are mainly consisting of linear patterns. Prominent among them are thick black bands (a salient feature of this phase), horizontal lines, zigzag or wavy lines and oblique or vertical strokes. The natural and geometrical designs are conspicuous by their complete absence. In regard to incision technique of decoration wavy or flowing finger marks, comb pattern, oblique nail impressions in registers, chevron pattern, etc., may be enumerated. The relatively higher frequency of incision designs is indicative of the general liking for this technique, perhaps in absence of required awareness about the range of painting designs, techniques and substances for this purpose. It also signifies both the incipient stage of the art of pottery painting and the imaginative/creative faculty of the pot-makers. This in turn scripts their utilitarian approach, which was adopted by the people in the given techno-socio-economic environment during that period of time.

In regard to shapes the overall scenario does not seem different in any manner. Because many typical shapes representing the A to E fabrics of early Harappan ceramic ware are not visible in circulation. The pottery shapes are limited to only certain types. These include bowls, basins, vases, troughs and lids, which are being discussed below. However, the pottery drawings are not to scale.

Description of Illustrated Pottery

Fig. 5.1: Early Harappan Pottery

1. Fragment of a vase in red ware with projected thick rim, concave neck, medium fabric. The painting is confined to portion above shoulders in the form of a thick black band over uneven red slipped surface, from Rindhana-II and Ghadwal-I.
2. Broken piece of a vase in red ware with short, thick projected rim, squat neck, expanding shoulders, of medium fabric. Painted in black forming a block black band around

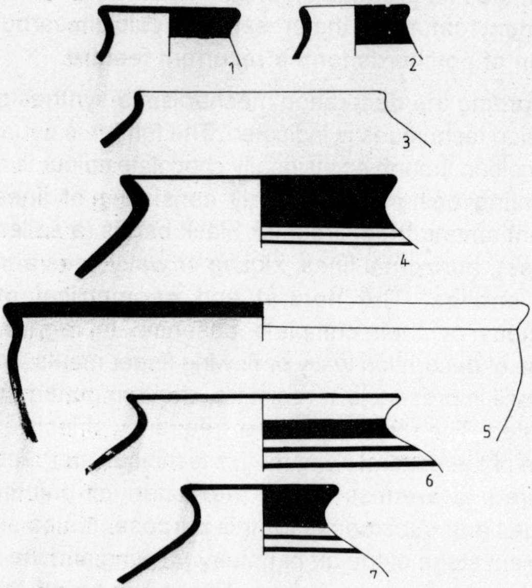

Fig. 5.1

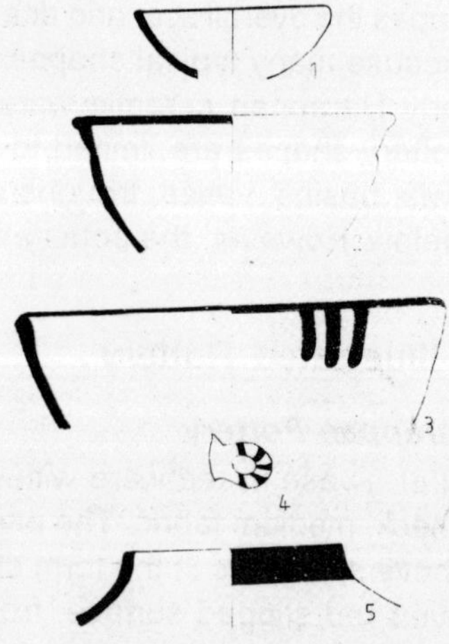

Fig. 5.2
Early Harappan Pottery

the rim, neck and shoulders. It was received from Rindhana-II, III, Gangesar-I and Silana.

3. This is a broken piece of a medium coarse fabric vase with brick red core, slightly out-turned short featureless rim, expanding shoulders and bulbous body. This comes from Rukhi-I, Gangesar-I, Ahulana-I and Baroda-III.
4. This is a fragmentary vase in red ware with concave neck approaching to carinated form, expanding shoulders, bulbous body and thick out-turned projected rim. It is an ill-fired specimen for the section is not uniformly red in colour as its core is slightly grey. A thick broad black band is witnessed around the rim and neck over plain red surface. Besides a narrow black band is also executed on the shoulders. This was collected from Ghadwal-I and Rindhana-II.
5. This is a broken piece of a basin of medium red fabric with a splayed out rim and round body. Its belly is rusticated like Kalibangan fabric-B. It was picked up from Silana and Gangesar-I.
6. This is a fragment of a vase with wide open mouth flaring rim, concave neck, tapering shoulders and bulbous body. This belongs to red ware, medium fabric and is painted in black over red surface. There is a fairly thick black band at the neck and two horizontal black bands at the shoulders are equally significant. It comes from Rindhana-III.
7. This is fragment of a vase of medium fabric with grey core—a sign of improper firing. Its mouth is reasonably wide, with out-turned projected rim, concave neck, tapering sides and bulbous round body. It is painted in black with a very broad band at the rim and neck along with two other horizontal thin black bands on the shoulders. This specimen comes from Rindhana-II and Ghadwal-II.

Fig. 5.2: Early Harappan Pottery

1. This is a broken piece of a bowl in red ware of medium fabric with light grey core—a sign of uncontrolled and imperfect firing technique. It bears convex sides, round

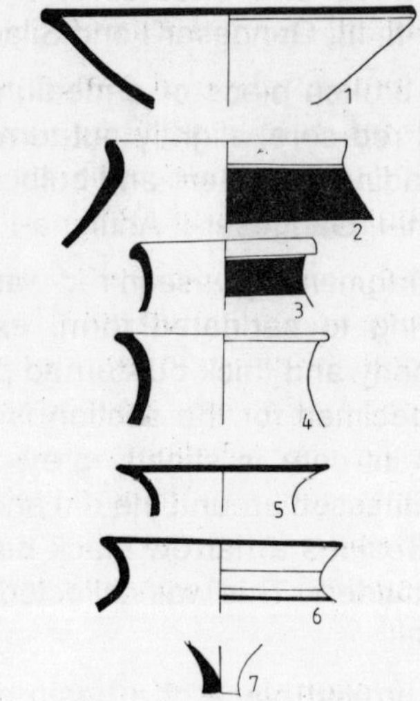

Fig. 5.3

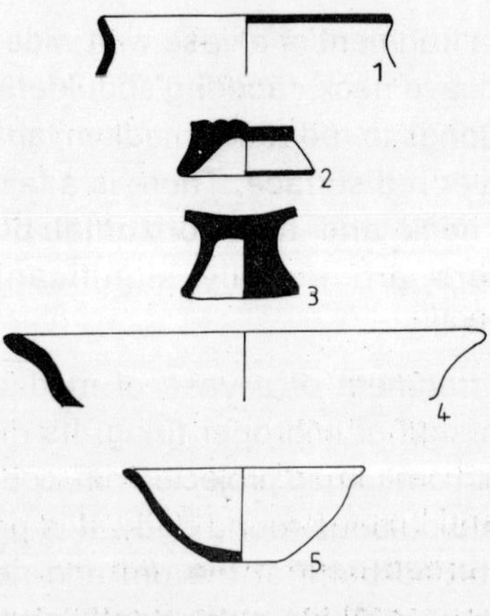

Fig. 5.4
Early Harappan Pottery

base and featureless flat rim. It was recovered from Rukhi I and Gangesar-I.

2. This is a fragment of a bowl in red ware of medium fabric, having slightly everted rim, tapering sides and round base. It is painted in black over red surface with a horizontal band at the rim. This was collected from Baroda-III.

3. This is a piece of a bowl in red ware of medium fabric with a featureless blunt flat rim and tapering sides. It is painted in black with three vertical strokes at the rim emanating from the main horizontal rim band. It was collected from Ghadwal-I.

4. This is a broken ring-shaped handle of red ware bearing black bands at regular intervals along the round section. This was recorded from Rindhana-II.

5. This is piece of a vase of medium fabric and grey core bearing featureless flat topped rim as if the art of making rim was not known to the potter or it is in its embryonic stage. Its mouth is considerably wide with squat neck, downward expanding shoulders and bulbous round body. It is painted in black with a broad band at the neck and rim over red surface. It belongs to fabric-A of Kalibangan-I and collected from Baroda-III.

Fig. 5.3: Early Harappan Pottery

1. Broken piece of a bowl with slightly flaring rim having black horizontal band around, medium fabric, dull red core, surface devoid of any slip or wash, from Rindhana-III and Gangesar-I.

2. Sherd of a vase in red ware with gently out-turned rim wide open mouth, concave squat neck, extending sides, medium fabric with red core, decorated with a block band in black covering rim, neck and shoulders, from Rindhana-III.

3. Piece of a vase in red ware with beaded rim, high concave neck, decorated with block black band over the plain red surface enveloping rim and neck, medium fabric, from Ghadwal-II.

4. Fragment of a vase with splayed out rim, high concave thick neck, extended profile, medium fabric in red ware and core, devoid of any decoration and surface treatment, from Gangesar-I.
5. Sherd of a vase with flaring rim, high concave neck of medium red ware with red core and pinkish red surface, from Baroda-III and Ahulana-I.
6. Fragment of a vase in red ware with flaring rim, concave short neck, expanding shoulders, medium fabric, red core, decorated with a thin horizontal band on the rim, a fairly thick black band at the neck and a thin black band at the shoulders, from Rindhana-III and Ghadwal-II.
7. Broken piece of a globlet like vase with hollow pedestal base of medium fabric pinkish red core and red surface, devoid of any slip, from Rukhi-I and Silana.

Fig. 5.4: Early Harappan Pottery

1. Fragment of a vase with gently out-turned rim, slightly concave small neck, extending profile, open mouth, medium fabric, red core, decorated with thin horizontal black band on the plain red rim, from Ahulana-I and Silana.
2. Piece of a vase with hollow and concave pedestal base having thin black band, medium fabric, red surface and core, from Gangesar-I, Silana, Rindhana-III and Baroda-III.
3. Fragment of a vase with concave and expanding sided pedestal base having a block black band around, medium fabric, greyish core and red surface, from Rindhana-III.
4. Sherd of a dish in red ware with projected rim, carinated shoulders, red core and plain red surface, devoid of any painting, from Gangesar-I and Silana.
5. Sherd of a bowl with everted rim, wide mouth, tapering sides, flat base, medium red fabric and pinkish red surface, from Baroda-III, Rukhi-I and Silana.
6. Sherd of a vase with everted rim, concave short neck, extending profile, open mouth, medium fabric, red core, decorated with horizontal band on the plain red rim, from Rindhana-III.

Harappan Pottery

Coming to the mature Harappan phase an allround development is observed in respect of pot–making, surface treatment, firing, decoration, etc. In this phase the pottery has been evolved to bichrome from polychrome as the designs are now executed only in black pigment over red slipped surface. The shapes have also been classified in the forms of goblet, beaker, 's'-shaped jar, perforated jar (fig. 5.72: 6), nail-headed dish-on-stand with prominent drums. As regards surface treatment red slip of fine quality is usually applied, giving it a very fair and attractive look. The pottery is made of well-levigated and prepared clay, which exhibits tremendous precision in firing technique as the core is usually brick-red producing a metalic sound on strike. The painting designs have been executed in dark black colour of fast nature (fig. 5.71: 1-5). And further standardization is attained in relation to painting designs. These include natural as well as geometric designs. Among them mention may be made of pipal leaf, fish scale, peacock, triangles, diamonds, etc. It is to be noted here in this case that all these characteristic features of the mature phase witnessed at the so far excavated sites in the vicinity like Banawali[7], Mitathal[8], and Balu[9] are not visible in relation to the explored settlements of this group in the area under discussion. But some of the features are recorded during exploration work. And the reason for their absence seems to have been partly their being the village settlements and partly the exploratory nature of this work. Nevertheless the very scant presence of such elements of the mature Harappan tradition is more than enough to attest human existence at some settlements in this area. These features will be dealt with in the following paragraphs to elucidate the gradual evolutionary process from early Harappan times to the mature phase.

Fig. 5.5: Harappan Pottery

1. Fragment of a dish with nail-headed rim and thick disc base, devoid of any surface treatment and any painting, red core, from Ghadwal-I.
2. Fragmentary shallow dish, red ware, medium fabric, short projected thick rim, from Rindhana-I.
3. Piece of a shallow dish-on-stand with slightly out-curved rim, from Gangana-II.

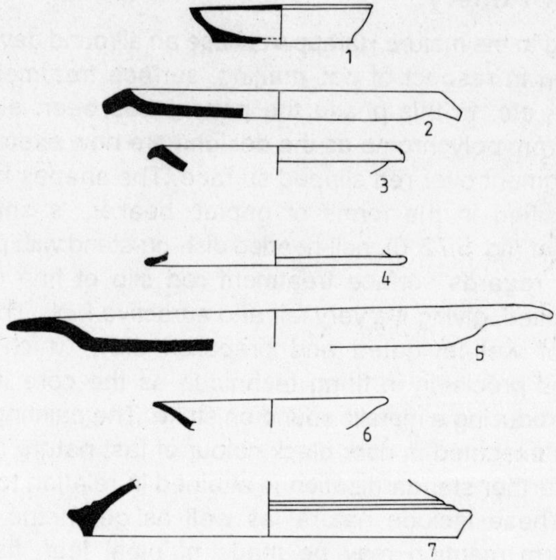

Fig. 5.5

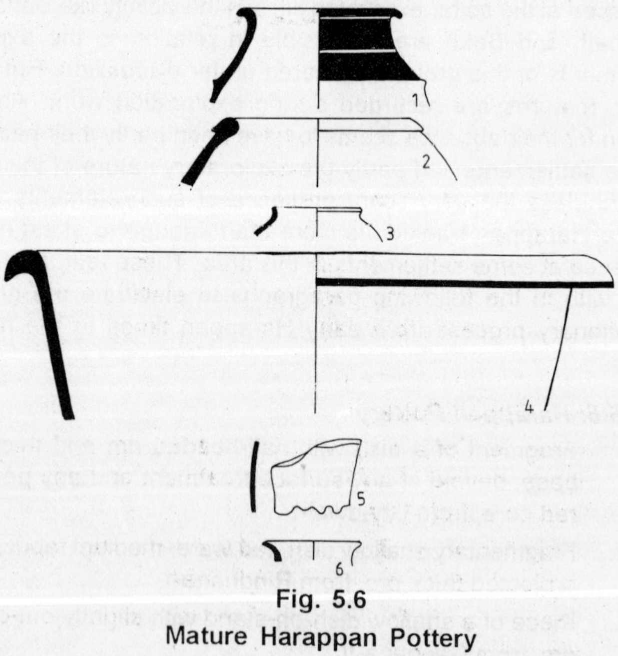

Fig. 5.6

Mature Harappan Pottery

4. This is a base part of a dish-on-stand with expanding lower sides assuming horizontal shape and forming a very broad-based stand from Ghadwal-II, Nuran Khera-I and Ahulana-I.
5. Fragment of a shallow dish-on-stand of red ware, medium fabric, devoid of any surface treatment, with large splayed out rim, mild carination at the shoulders and flat wide base, from Baroda-III and Rindhana-I.
6. Fragmentary dish-on-stand with short projected rim, tapering sides and flat base, medium fabric and pinkish core, from Gangesar-I.
7. Broken lower portion of a stand of dish with downward expanding sides, taking shape of a broad–based stand with multiple ridges near the lower end resulting in a short projection, thick, red ware, treated with fine slip, brick-red core, from Ahulana-I.

Fig. 5.6: Harappan Pottery

1. This is a piece of a red ware vase with out-curved undercut rim, ridge at the shoulders, tapering sides, painted in black with broad block band over rim, neck and shoulders, a remnant of early Harappan tradition, from southern cuttings at Rindhana-I.
2. Fragment of a vase with wide open mouth, globular body, externally downward bevelled or collared rim, medium fabric and pinkish red core, treated with light red slip of worn off nature, from Rindhana-III.
3. Piece of a vase of red ware of medium fabric, red core, out-turned bevelled rim, concave short neck, expanding sides taking shape of globular body, from Baroda-III and Kahalapa-II.
4. Sherd of a storage jar in red ware, thick fabric, beaked rim, gradually tapering sides, decorated with black band at the rim, from the southern levelled portion at Ghadwal-III.
5. Broken piece of a perforated jar in medium red ware with brick-red core and red slip, from the north-western levelled periphery of Ghadwal-I.

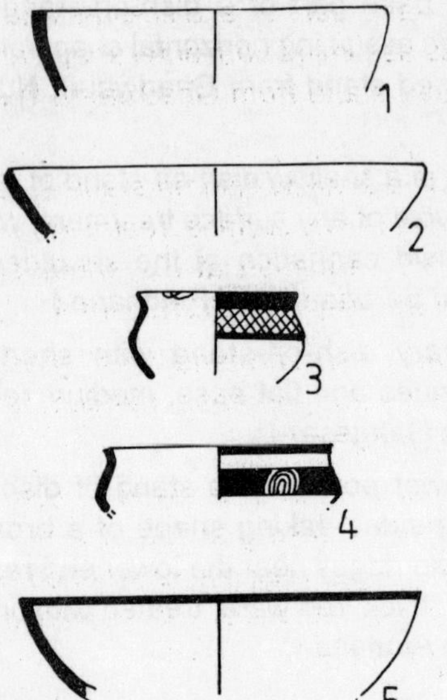

Fig. 5.7

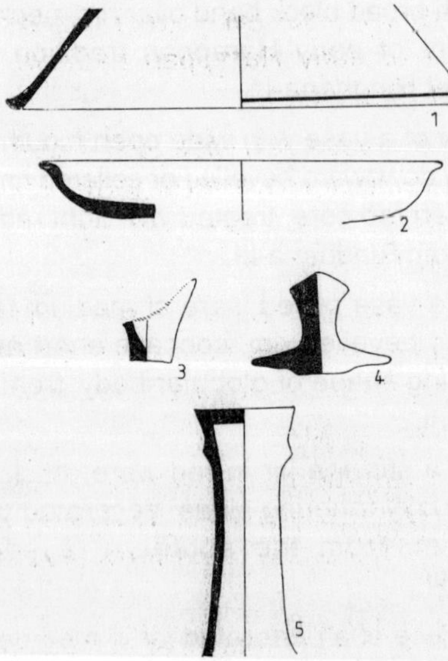

Fig. 5.8

Mature Harappan Pottery

6. Fragmentary vase with flaring rim, concave neck and expanding shoulders, medium red fabric, red core, treated with red slip of fugitive nature from Baroda-III.

Fig. 5.7: Harappan Pottery

1. Broken piece of a bowl with featureless flat-topped rim, convex sides, devoid of any decoration, medium fabric and red core, from the cuttings at Bhawar-I.
2. Fragment of a bowl in red ware, medium fabric, red core, decorated with thin black band at the rim over red slip. It is with featureless rim and convex sides, from Ahulana-I and Gangesar-I.
3. Fragment of a small bowl with everted short rim, open mouth, tapering sides forming a carination at the belly, round base, medium fabric, red ware and core, cross-hatched painting design, from Gangesar-I cf. Mitathal IA fig. 7: 20.
4. Sherd of a bowl with slightly everted rim, wide open mouth, concave short neck, downward tapering sides, carinated at the belly, round base, medium red fabric and core, decorated with a black band and loop design, from Kahalapa-II, cf. Mitathal IIA Fig. 7: 21.
5. Fragment of a shallow bowl with wide open mouth, flat-topped featureless rim, downward receding sides, round base, medium fabric, red ware and core, decorated with a black band on rim, from Bhawad-II.

Fig. 5.8: Harappan Pottery

1. Fragmentary base of a dish-on-stand, of medium fabric, brick red core, treated with red slip over which black horizontal band is executed near the ridge, from Rindhana-I and Ghadwal-I.
2. Fragment of a dish-on-stand with projected rim, round sides, flat base, medium fabric, brick-red core and treated with a fine red slip, from Ghadwal-I, Baroda-I and Gangana II cf. Mitathal II A Fig. 6: 13.
3. Narrow flat base of a vase in red ware, medium fabric, red core and treated with red slip, from Nuran Khera-I.

4. Fragment of a lid in red ware with central conical knob, medium fabric, red core and plain surface, from Bhawar-II.
5. Sherd of a large stem of a dish-on-stand in red ware with a prominent sharp rib, medium fabric, brick-red core and treated with red slip on the outer surface, from Ghadwal-I.

Late Harappan Pottery

In the succeeding phase popularly known as Late Harappan, a distinct process than the preceding one is to be marked, which signifies onset of a devolutionary process. Although the pottery is still wheel made and of bichrome category, yet, the typical features of the mature Harappan class are conspicuous by their absence. In this phase clay is not well-levigated as indicated by foreign elements in the form of small pieces of calcium carbonate and husk in the sections of potsherds. In regard to surface treatment the fine finishing touch of the preceding phase giving lustrous look to the surface is lacking. And the application of red slip is not common feature now, which also lacks in uniformity and quality. This trend is further testified in painting designs and shapes also. The painting designs seem to be growingly reduced to linear and repetitive types. The variety, richness and precision of the yester-years are seen giving way to simple linear forms. Moreover an increasing application of incision technique for decoration is becoming the order of the day. The characteristic features of the late Harappan pottery are being viewed in the light of the richness of mature Harappan pottery as such irrespective of any consideration of a site(s) location, whether it belongs to the area under consideration or from some other site. These signify a shift in their thinking process, artistic ability to execute their creative ideas in sophisticated manner and the changing socio-economic conditions. This also highlights a change in their outlook. As a result dish-on-stand with drooping blunt rim takes place of dish-on-stand with nail-headed rim. And similarly shapes like deep dishes, basins, bowls, vases and jars with beaded or collared rim become popular.

It seems that in this respect a resurgence of early Harappan traits of Siswal and Sothi village cultures is intermixed with some of the devolved Harappan traits. The ceramic ware of this phase registers a fine admixture of these traditions. However, the ceramic industry of this phase is consisted of a dull red ware of medium to sturdy fabric of soft firing technique. Occasionally its exterior tend to peeloff giving a look of the Ochre Coloured Ware. However, the ceramic industry primarily belong to the Harappan group for its close family affinity. A brief survey of the main features is as follows:

Fig. 5.9: Late Harappan Pottery

1. Fragment of a dish-on-stand, red ware, medium fabric, red core, short projected undercut hooked and beaded rim, carinated at the shoulder, tapering sides and round base, treated with red slip of fugitive nature, without any painting. This is a popular form and comes from Ghadwal-II and III, Bichpari-I, Kathura, Ahulana-II, Isapur Kheri, Kahalapa-I and Rindhana-IV.
2. Broken piece of a dish-on-stand with drooping, blunt dull rim, downward sharply receding sides, decorated with black band and a leaf like design filled with vertical lines. This is also a type shape of this period, comes from Rindhana-IV, Jawahara-I, Kakana and Baroda-III.
3. Sherd of a dish-on-stand with thick, blunt drooping rim, thick tapering sides and painted in black over red surface with registers of opposite oblique lines forming blank triangles in between, from Gangana-I, Busana, Rindhana-IV and Ghadwal-II.
4. Fragment of a dish-on-stand with short drooping dull rim and ridge on the stand near the base of dish, thick sturdy fabric with red core and slip, painted in black with a wheat-seed tube or leaf like design enclosed within thin rim bands, from Rindhana-IV.
5. Fragment of a dish-on-stand with thick projected, dull and blunt rim, round sides, medium fabric, red ware and core, from Jawahara-I.

Fig. 5.10: Late Harappan Pottery

1. Broken piece of a red ware dish with slightly projected thick and dull rim, pinkish core, untreated dull red surface, from Bhawad-I, Rukhi-II, Butana-I and II.

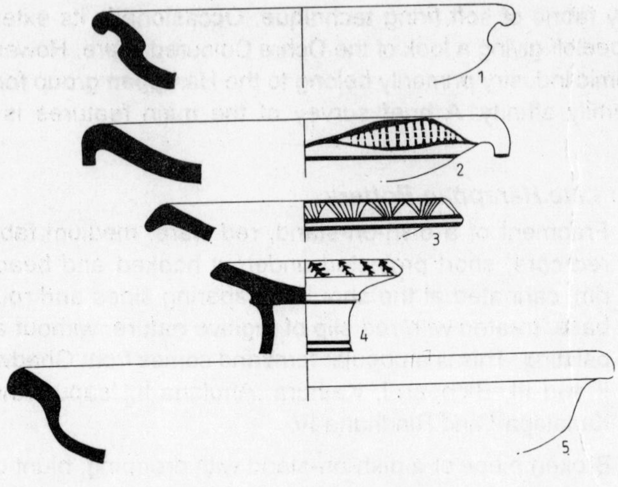

Fig. 5.9

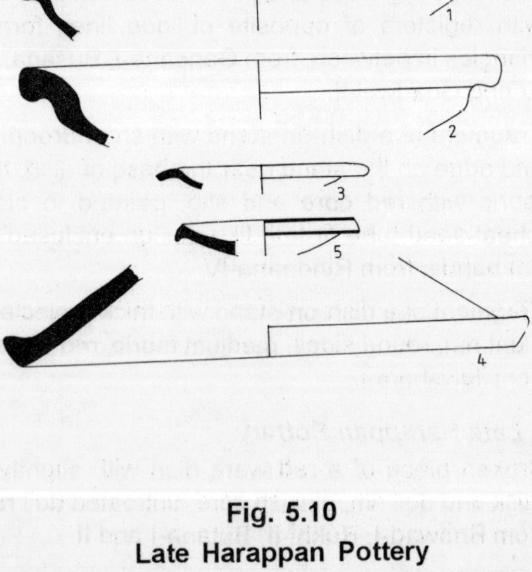

Fig. 5.10

Late Harappan Pottery

2. Fragment of a dish-on-stand with thick, dull and projected rim, slightly carinated shoulders, relatively thin tapering sides taking shape of round base, medium fabric, red core and unattractive red surface, from Nuran Khera-II and Gangesar-I.
3. Sherd of a medium fabric, red ware dish with projected blunt rim, tapering sides, treated with dull red slip and devoid of any painting, from Kahalapa-I.
4. Fragment of a dish stand with thick, sturdy and upturned wide base, medium fabric, light red surface and core, from Sisana-I.
5. Broken piece of a dish-on-stand with short drooping rim, plain red surface, thickened sides near the centre and medium fabric, from Ghadwal-I.

Fig. 5.11: Late Harappan Pottery

1. Broken piece of a red ware basin with projecting thick rim and round sides, medium fabric, pinkish red core and surface, from Laharada-I and Bohla.
2. Fragment of a basin in red ware with projecting sides, beaked small rim, wide open mouth, flat base, medium red fabric, treated with red slip, from Butana-I and Bali-I.
3. Fragment of a basin with everted rim, open mouth, tappering sides, roundish base, medium fabric, grey core, painted with a black band on the rim, from Busana Kalan, Jawahara-I and Sisana-I.
4. Broken sherd of a basin with short flaring undercut rim, wide open mouth, roundish body, red ware, medium fabric, greyish core, decorated with thin black band over red slip, from Gohana-I.
5. Dish-on-stand with hooked, undercut, thickened rim, tappering sides, round base, thick sturdy fabric, dull red surface and core, from Laharada-I.

Fig. 5.12: Late Harappan Pottery

1. Bowl piece with beaked rim, round profile, medium fabric, red core and ware, worn out surface, from Saragthal.
2. Fragmentary bowl in red ware, medium fabric, with splayed

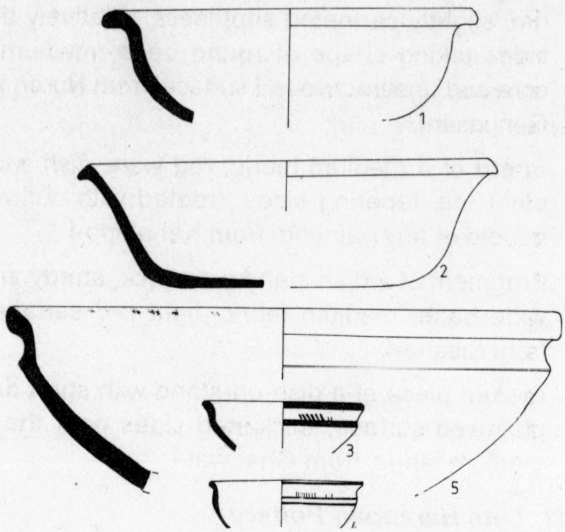

Fig. 5.11

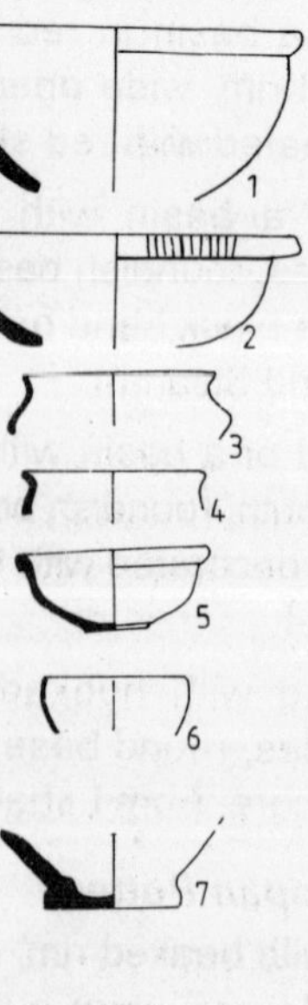

Fig. 5.12
Late Harappan Pottery

out rim, round sides, flat base, plain red surface comes, from Banwasa.

3. Fragment of a bowl with everted rim, concave neck, sharp carinated shoulders and tappering profile, plain red ware, medium fabric and pinkish red core, from Ghadwal-III.
4. Broken piece of a bowl with slightly everted rim, convex sides, pinkish red surface, greyish core, medium red fabric, from Rukhi-II, Ahulana-II and Busana Kalan.
5. Sherd of a bowl, medium red fabric, pinkish core, thin red slip, beaded rim, tapering profile and disc base, from Gangesar-I.
6. Fragment of a bowl with flat topped featureless rim, convex sides, medium fabric, red core and plain red surface, from Busana Kalan.
7. Broken piece of a button based vase, red ware and surface, medium fabric and greyish core, from Ahulana-II.

Fig. 5.13: Late Harappan Pottery

1. Fragment of a jar with beaded rim, squat neck, tappering sides forming a bulbous body, red ware, medium fabric, coated with dull red slip of fugitive nature, from Rindhana-IV and Gangesar-I.
2. Broken piece of a jar with collared rim, extended profile approaching to bulbous shape, red ware of medium fabric and coated with dull red slip, from Ahulana-II.
3. Fragment of a vase with beaded rim, concave high neck, expanding sides, dull red surface, red ware, medium fabric, decorated with black horizontal band on the rim, from Pipli-I.
4. Broken piece of a vase with beaded or beaked rim, medium sized concave neck, medium fabric in red ware, from Isapur Kheri.
5. Sherd of a ring stand with out-turned rim and slightly in-turned flat base, medium fabric, red ware and dull red surface, from Rindhana-IV.

Fig. 5.14: Late Harappan Pottery

1. Fragment of a red ware jar with collared rim, wide open

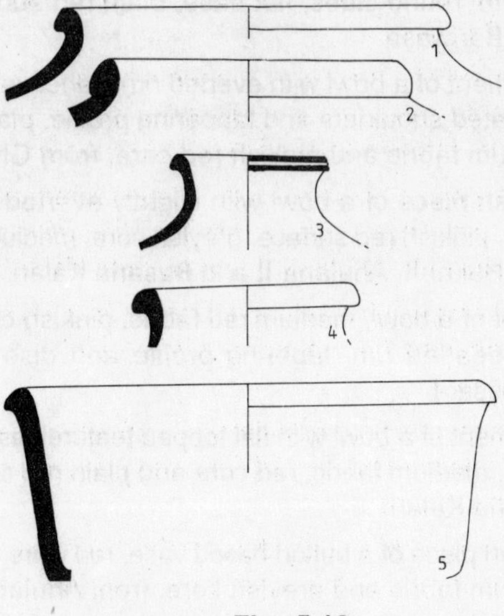

Fig. 5.13

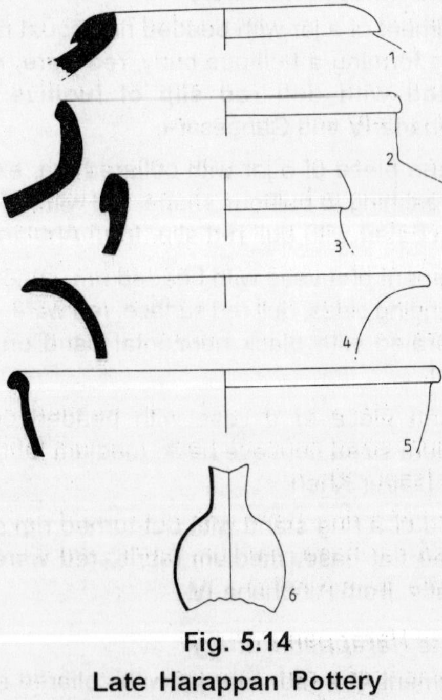

Fig. 5.14

Late Harappan Pottery

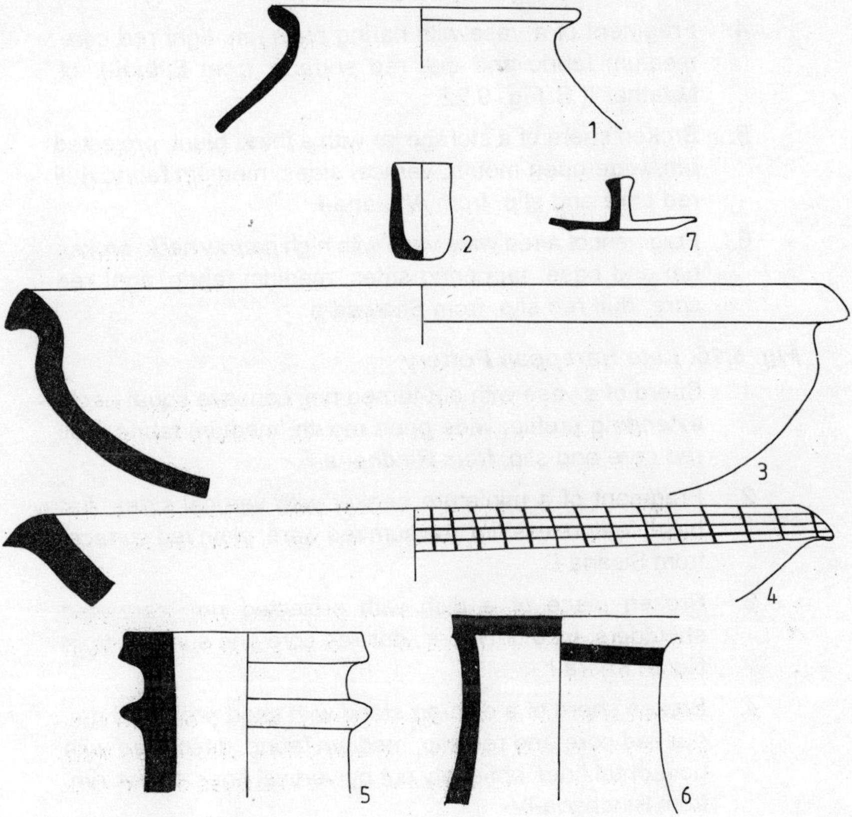

Fig. 5.15 : Late Harappan Pottery

mouth, extending sides, medium fabric, dull red slip, from Pipli-I, cf. Mitathal II B Fig. 8:6.

2. Broken piece of a storage jar with slightly everted rim, almost vertically straight neck, expanding profile adopting a bulbous shape, medium red ware, from Ahulana-III.
3. Sherd of a vase with beaked rim, slightly concave neck, extending shoulders, medium fabric, red core and coated with dull red slip, from Banwasa.
4. Fragment of a vase with flaring plain rim, light red core, medium fabric and dull red surface, from Bidhal-I. cf. Mitathal II B Fig. 9:22.
5. Broken sherd of a storage jar with a thick, blunt, projected rim, wide open mouth, vertical sides, medium fabric dull red core and slip, from Ahulana-I.
6. Fragment of a red ware vase with high narrow neck, broken rim and base, tappering sides, medium fabric, light red core, dull red slip, from Bhawad-II.

Fig. 5.15: Late Harappan Pottery

1. Sherd of a vase with out-turned rim, concave squat neck, extending profile, wide open mouth, medium fabric, dull red core and slip, from Rindhana-I.
2. Fragment of a miniature beaker with vertical sides, flat base, featureless rim, medium red ware, plain red surface, from Sisana-I.
3. Broken piece of a dish with projected rim, carinated shoulders, medium fabric, dull red core and surface, from Nuran Khera-I.
4. Broken sherd of a dish-on-stand with short projected rim, dull red core and red slip, medium fabric, decorated with horizontal lines obliquely cut by vertical lines on the rim, from Rindhana-IV.
5. Fragment of a stem of dish-on-stand, medium fabric, red ware with two sharp ribs below the junction, red core, light red slip of fugitive nature, from Rindhana-IV.
6. Broken stem of a dish-on-stand with a black horizontal band below the junction, medium fabric, red core and slip, from Ghadwal-III.

7. Broken lid in red ware with a central flat topped vertical knob, flaring sides, flat base, medium fabric, red surface, from Bidhal-I and Jawahara-I.

Painted Grey Ware

We witness appearance of a new ceramic ware, popularly known as the Painted Grey Ware, in the subsequent phase. This is a novel innovation in the field of ceramic industry and as many as 39 settlements have produced remains of this ware from the district under consideration besides 34 Grey Ware sites (figs. 5.38-41). Going by the general observation regarding frequency of this ware at the explored and excavated settlements in this district and elsewhere respectively it seems that the percentage of this deluxe ware is not more than 10 to 15 per cent of the total ceramic assemblage. The name is assigned mainly because of the novel experience, though the dominating ceramic industry is still red ware. The main characteristic features of this ware are the following ones. The pottery is wheel thrown, made of well-levigated clay, with thin to medium section (a sign of remarkably controlled firing technique) uniformly applied grey slip either on one or on both the sides gives it further a very attractive look. It is painted in jet black colour of very fast nature and the paintings have been executed on both the sides. The paintings are consisting of linear, geometrical and floral designs. Among them mention may be made of horizontal and vertical lines, dots, circles, intersecting circles, four petalled flowers, etc. At times incision technique have also been seen adopted for decoration purpose. Thus, one can conclude that on the whole the pottery decoration is very interesting. As regards the pottery shapes it can be stated that dishes, basins, bowls, vases and miniature vessels of various shapes with varied types of rims are especially noteworthy. Lota-shaped vase is the Characteristic type of this ware. So far as the comparative study of the pottery of late Harappan and the succeeding PGW levels is concerned an overall improvement in case of the latter is quite obvious. However, the other industry of this period is plain red ware mainly of coarse fabric comprising shapes like jars, vases and bowls (fig. 5.42). A detailed study of these and the late Harappan potteries from excavated sites is called for to underline the element of continuity and change, if any, during the period. However the main pottery shapes and painting designs of PGW levels are to be discussed in the following lines.

Fig. 5.16: Painted Grey Ware

1. Dish of grey ware with slightly incurved sharpened rim, convex sides, devoid of any painting, thin section, well fired and slipped, from Harsana Kalan-I and Jawahara-I.
2. Dish of grey ware, thin fabric and incurved sharpened rim having black bands on both sides of the rim, convex sides and round base, from Jawahara-I.
3. Dish of Painted Grey Ware, medium fabric, well-fired, incurved featureless rim, convex sides, round base and decorated with black dots, from Ghilaud Khurd, Rukhi-II and Gangesar-II.
4. Dish with incurved sharpened rim, convex sides, round base, medium fabric, slipped and painted with vertical strokes emanating from black rim band on both the sides, from Rindhana-V.
5. Dish with incurved featureless rim, convex sides, round base, devoid of any painting, but bearing slip. It is well-fired to produce uniform colour, from Bhawad-I, Nuran Khera-II and Jawahara-I.

Fig. 5.17: Painted Grey Ware

1. Fragment of a basin with sharp collared rim, slightly convex sides, medium fabric, devoid of any painting design, grey core-sign of proper and controlled firing, gathered from Chatia-deva, Ghilaud Khurd and Datauli.
2. Broken piece of a basin with externally grooved rim, convex sides, medium fabric of well-fired grey ware, from Sisana-II and Rithal.
3. Fragment of a basin with vertical, flat topped rim, convex sides, medium fabric, devoid of any painting and slip, of fine grey ware, from Jawahari-I.
4. Basin with thick collared rim with two grooves below, medium fabric, devoid of any painting and slip, from Kakana, Thuska and Bohla.

Fig. 5.18: Painted Grey Ware

1. Broken sherd of a bowl with vertical sharpened rim, straight sides with carinated sagger base and medium fabric of

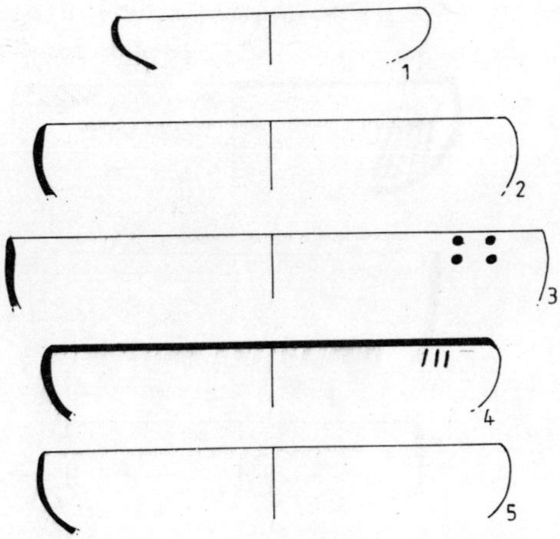

Fig. 5.16

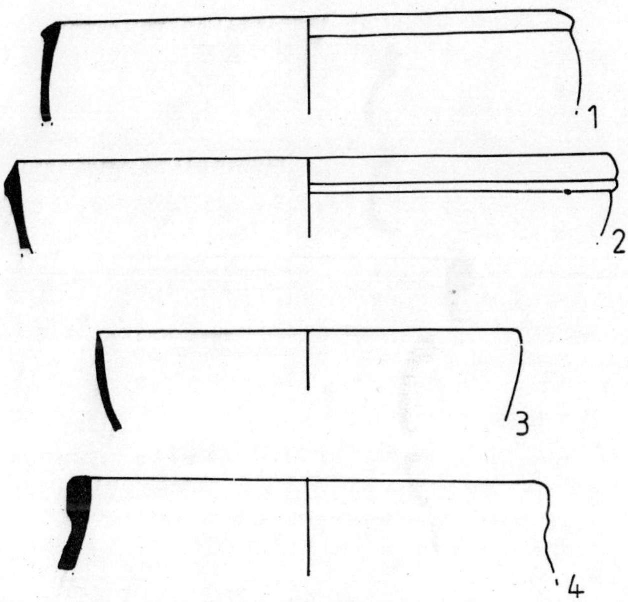

Fig. 5.17
Painted Grey Ware

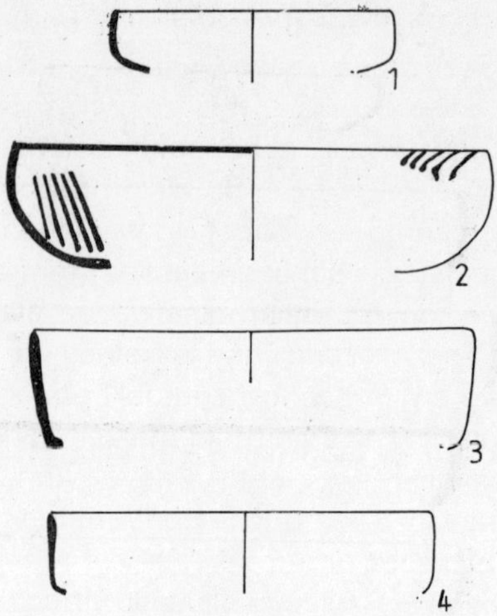

Fig. 5.18

Painted Grey Ware

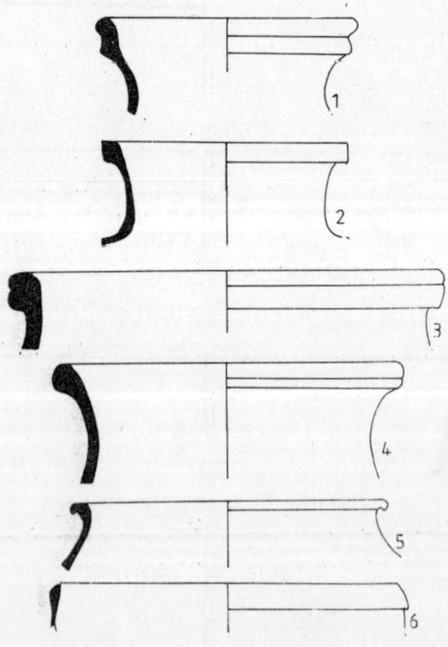

Fig. 5.19

Grey Ware and Associated Red Ware

plain grey ware, from Nangal Khurd, Jawahari-I and Sandhal Khurd.

2. Bowl with an incurved sharpened rim, convex sides, round base, medium fabric, painted with a thin rim band on both sides, from Akbarpur Barota and Harsana Kalan-I.
3. Sherd of a bowl of dark grey ware, with vertical blunt rim, straight sides with carination at the sagger base, without any slip and painting, from Panchi Gujjaran and Pipli Khera.
4. Bowl with flat featureless rim, vertical sides, sagger base and medium fabric of black-slipped ware, from Gumad and Akbarpur Barota.

Fig. 5.19: Grey Ware and Associated Red Ware

1. Fragment of a red ware vase with out-turned double-ribbed rim, medium sized concave neck, medium fabric, plain red surface, from Khanpur Kalan-I and Harsana Kalan-I.
2. Broken piece of a red ware vase, round thick fabric, grey core, vertical neck, decorated with multiple parallel grooves on the rim, from Sisana-II.
3. Sherd of a red ware vase with an out-curved thickened and externally grooved rim, dull red fabric, core and outer surface, from Barota and Gumad.
4. Fragmentary sherd of a vase in red ware with an out-curved thickened and externally grooved rim, medium fabric, grey core and plain dull looking outer surface, from Sisana-II.
5. Broken piece of a grey/red ware vase with an out-turned flat topped and undercut rim, rough fabric, from Jawahari-I.
6. Sherd of a deep bowl of grey ware with an in-turned and externally clubbed rim having slip on both the sides, from Nakloi-I.

Red Ware–NBPW

After the Painted Grey Ware levels, in the succeeding period, a resurgence of Red Ware group of pottery is witnessed in the time period of Northern Black Polished Ware in the early historical period. The pottery in this phase is of medium coarse fabric with prominent signs of soft firing process, reducing the section usually dull red to grey shade. The outer surface of the pottery is also not usually smooth. It is dull red in colour with grey core, in agreement with the firing technique. This also signifies the growing precarious states of the art of potting and surface treatment. Similarly regarding the art of decoration and repertoire of painting designs the prevalent state does not appear inspiring in any manner. This is amply displayed by currency of only a few horizontal incised grooves and stamped designs over the light red-washed outer surface. And almost complete absence of paintings. This speaks of the devolutionary process of the art of pot-making in so far as the red ware is concerned or should these be taken as a sign of their utilitarian significance? Pottery shapes are also restricted to simple bowls, basins and vases (fig. 5.42: 2, 4, 16). In regard to NBPW one can only suggest that since the ware was recorded from a few sites and that too in the form of small broken pieces it was scarce in this area. This restricts our analysis to only the red ware group which are elaborated in the following lines.

Fig. 5.20: Red Ware–NBPW

1. Fragment of a vase with everted, but obliquely outward bevelled rim, concave medium sized neck, medium fabric and worn out surface, from Ibrahimpur and Harsana Kalan-II.
2. Sherd of a vase with out-turned blunt, thickened rim having multiple groove in a chain on the outer surface of concave neck, expanding sides, medium red fabric, dull red core with greyish tinge and plain red surface, from Datauli and Panchi Gujjaran.
3. Fragment of a jar with flanged rim, concave neck with gentle multiple grooves on the outer side, expanding shoulders and sides promising a globular shape of medium rough fabric and dull red core, from Bali-II, Khanpur Kalan-II and Ghilaud Kalan settlements.
4. Fragment of a vase with projected/splayed out rim, concave neck with a series of grooves on the external side, extending profile, of medium fabric, from Jawahari-II and Nahari-II.

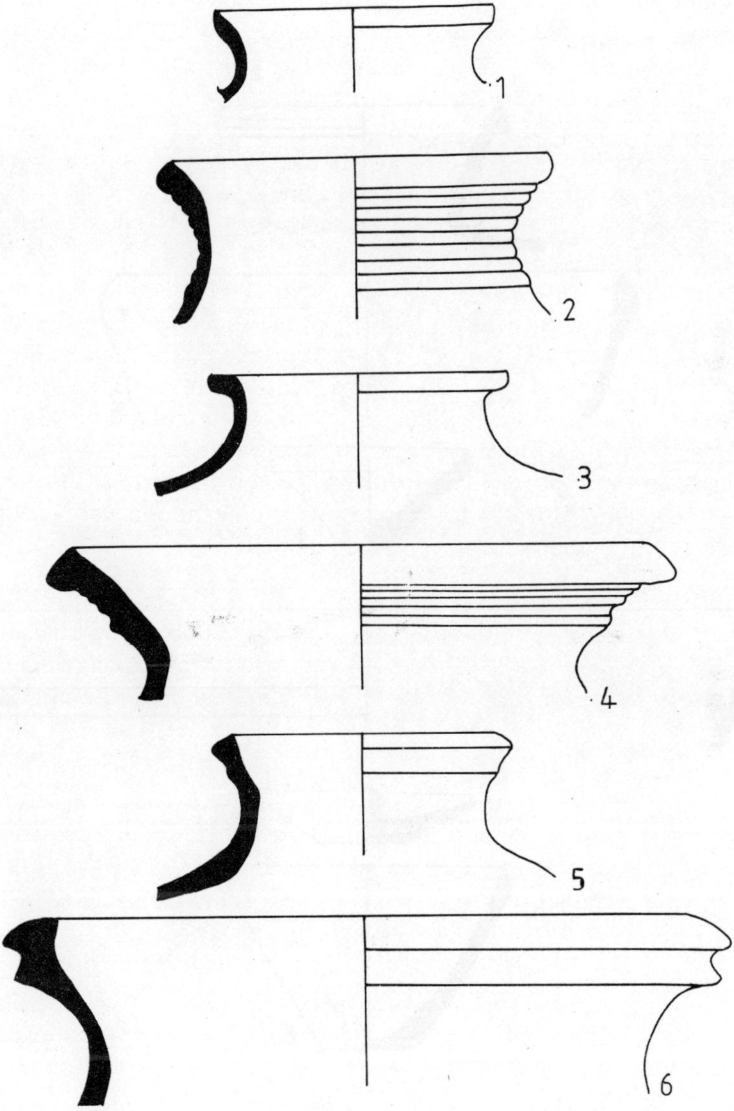

Fig. 5.20: Red Ware-NBPW

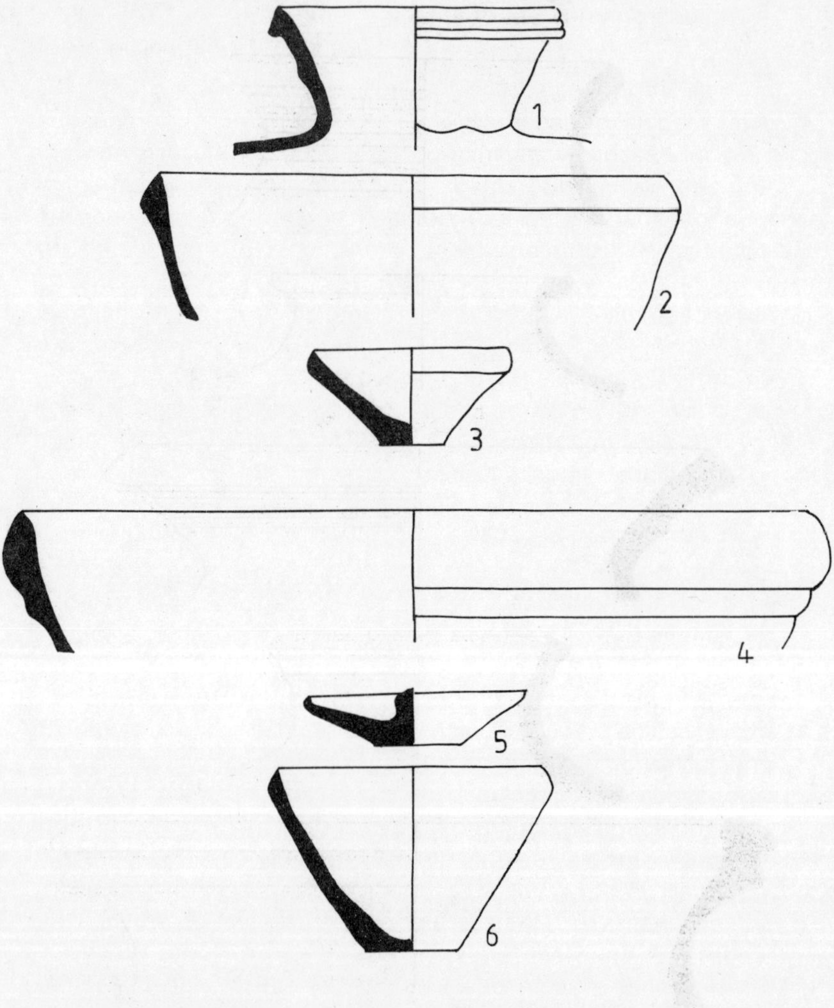

Fig. 5.21: Red Ware-NBPW

5. Piece of a vase with everted and outward obliquely bevelled ribbed rim and concave medium sized neck, expanding shoulders assuming a globular shape, of coarse red fabric and worn out red surface, from Mahmoodpur and Jagasi.
6. Fragment of a vase with an everted flattened top and prominently ribbed rim, concave neck and extending profile, of medium fabric and red surface, from Thuska, Rindhana VI and Rukhi-III.

Fig. 5.21: Red Ware–NBPW

1. Fragment of a vase with a thick out-turned and externally grooved rim, narrow squat neck, medium fabric, burnt red core, dull red surface with signs of red slip, from Badwasani and Thana Khurd.
2. Sherd of a bowl with sharp vertical and externally oblique bevelled rim, tapering sides, medium fabric, greyish core, treated with red slip, from Jasarana.
3. Broken piece of a bowl with in-turned and externally bevelled rim, receding sides, flat solid base, medium fabric, dull red core and worn out surface, from Rukhi-III.
4. Fragment of a basin with almost flat topped, blunt, round collared and externally grooved rim, medium fabric, dull red/grey core and surface, from Mahmoodpur and Pipli-II.
5. Sherd of a lid with conical central knob, flat base, projected sides, medium fabric, grey core and plain red surface, from Nakloi-II.
6. Fragment of a bowl with vertical sharpened and externally oblique bevelled rim, sharply receding sides, flat base, medium fabric, grey core and plain red surface, from Mandori.

Pottery Decoration Techniques

The history of pottery decoration travels back to the very inception of ceramic art. It, of course, experienced various vicissitudes over the period in accordance with the changing fortunes of the ceramic industry, which in turn is a reflection of the overall level of development of society. In the present context the beginning of the decoration process is closely associated with the very first phase, early Harappan of the Indus Civilization, which is not only the formative stage of the Harappa Civilization itself, but also of the great ceramic tradition, bichrome ware, of this period and thereby the starting point for the present discourse as well. The decoration is usually comprising of two techniques, namely, the incision and painting. With the help of these two methods people always attempted to execute their ideas into practice by way of either adopting natural or geometrical objects. In a way these two categories of items formed the subject matter of expression of their ideas through decoration of pottery types. These are to be taken up in the following lines with regard to various stages of development during the period under discussion.

In the early Harappan phase people adopted white, chocolate and black colours to decorate their ceramic ware. These colours are applied either on the plain red or red slipped/washed surface. As a result the pottery of this phase is popularly known as 'polychrome ware'. Specimen of this class were picked up, though very limited, from the surface of the sites representing this phase. Some of them, however, bearing signs of this colour combination have been shown (Figs. 5.22:1, 4, 7; 24: 4; 25: 2, 9). Besides there is an example of a combination of chocolate and brown colours over red surface (Fig. 5.23: 6). Thus, the paintings were executed in the aforesaid colours, which turned it into a polychrome ware.

In regard to the painting repertoire we may mention here the application of black block band over the surface area comprising neck, rim and shoulders (Fig. 5.22: 2-5, 7-9) cf. Mitathal-I,[10] figure, 5:1, 4-6, Mahorana,[11] figure 1:1-3, 5, Nayabans,[12] figure 2: 1-4, Khatoli,[13] figure 1:14, Sothi,[14] figure 53.3:3,20, Nohar[15], figure 53.4:4,10,12; Sherpura,[16] figure 53. 5:1,2,4,5,8,9, Dhermajra,[17] figure 30. 10: 24, 27, 30, 39, Hulas[18] figure 30.9:12; Alamgirpur,[19] figure 30.7:10,12,13; black bands (Figs. 5.22:1,4; 23:1,2,4; 24:8; 25:1,8); registers of black lines intersecting diagonally to form either triangles in the middle on fabric-A pottery (Fig. 5.22:5) cf Siswal[20], PL. XXXIV: 5, Sanghol[21], figure 30.13: 25; oblique strokes in a panel stemming from a horizontal black band (Fig. 5.22:4) on fabric-A sherds cf Siswal-A[22], PL XXXIV: 2,8, Hulas[23] 113:8; black vertical lines in a group of two and more (Fig. 5.25:6) on fabric-C sherd cf Siswal[24], PL. VI: 1; registers of black flowing lines in a reserved slip manner (Figs. 5.23:3-5; 24: 7-9) on fabric-C pottery cf Rakhigarhi[25], PL. XXVII: 5, Sherpura[26] 53.5:30, Ganeshwar[27], figure 72; a panel of black thin horizontal lines on fabric-C sherds (Figs. 5.23:7; 24:11) cf Siswal[28]; PL. XXXIV: 13. Likewise there are some examples of incision decoration which also deserve mention here. In this connection incised finger marks on the interior of dark grey colour pottery representing fabric-E and F of Kalibangan-I ceramic tradition (Fig. 5.27:1-2) cf Rakhishahpur[29]; horizontal lines (Figs. 5.28:2-5,7; 30:1-3) on fabric-B sherds cf Mitathal-I[30] PL. VI: 10; flowing lines on fabric-D sherds (Fig. 5.30:3-4,8) cf Mitathal-I[31], PL. VI:3; and irregular lines on fabric-D sherds (Fig. 5.28:12); finger marks on the interior (Fig. 5.29:5); flowing finger marks on the interior and rusticated exterior (Fig. 5.28:1) cf Mitathal-I[32], PL. VI:8; incised cord marks (Figs. 5.27:5; 28:1) on fabric-D pottery cf. Mitathal-I[33], PL. VI: 4, 9; incised shallow horizontal lines on the interior and rusticated exterior (Fig. 5.27:3) on fabric-D sherds cf Mitathal-I[34], PL. VI:5; externally rusticated surfaces as found in fabric-B in the manner of wet wares of Baluchistan (Fig. 5.27:8; 28:8) cf Rakhishahpur-I[35], PL. XXVII: 15 are very significant to suggest circulation of these decoration designs during this phase in this area. This also points to currency of Kalibangan-I fabrics, though surely in a restricted manner as only a few such sherds have been met with, among the early settlers of this area. A miniature vase with button disc base, round sides and narrow rising broken neck in dark grey colour deserve mention (Fig. 5.26).

A few instances of mature Harappan painting designs speak of decoration system of the succeeding phase. These are executed in fast black colour as usual over red/pinkish red/pinkish pale slipped surfaces (Fig. 5.71). The slip is always very fine giving the surface a lustrous look. The painting designs are mainly consisting of a few simple examples of peacock (Fig. 71:2), rhomb (Fig. 5.71:1,4) and fronds (Fig. 5.71:3) like parallel horizontal black bands around neck, shoulders and belly of vases of varying thickness and sizes (Figs. 5.24:1-3; 71:2); black flowing lines forming a chain of opposite small triangles enclosed by black bands (Fig. 5.25:10) cf Mitathal-II A[36], PL. VIII: 1,5, Alamgirpur[37], figure 107: 11; Cross-hatched diamonds over pale red slip on a medium ware sherd from Rindhana-III (Fig. 5.24:5) cf Mitathal-II A[38], PL. VIII: 11; black horizontal bands enclosing criss-cross design over pinkish slip on a thin red ware sherd from Ghadwal-II (Fig. 5.24:10) cf Mitathal-IIA[39], PL. VIII: 10, Hulas[40], figure 113: 6, Alamgirpur[41], figure 107: 2,10, Bara[42], figure 104: 2; Banawali[43], figure 61: 6; parallel horizontal bands enclosing band of incised wavy lines in the reserved slip manner (Fig. 5.24:2) cf Mitathal-IIA[44], PL. VIII: 2 and a maltese-cross hatched with oblique lines and multiple loops/ circles enclosed by multiple horizontal bands around neck (Fig. 5.25:3). Such malte-scrosses are available right from early Harappan to PGW phase as referred to by Bala.[45] But interestingly none of the crosses bears close affinity with the one in question. However, some resemblance can be established to the one recorded from Mitathal-II B[46], figure 9:35. With regard to the mature Harappan pottery one can also take note of a few sherds of perforated pottery (Fig. 5.31:1-3; 72:6), which were picked up from Chhapra, Gangesar-I and Ghadwal-I respectively. The ceramic complex is thus, in agreement with the general Harappan scenario witnessed in the region on the whole. All these examples underline continued occupation of this area by the Harappan people after the early Harappans.

In the succeeding late Harappan phase the ceramic ware betrays a family likeness with the preceding mature Harappan phase in respect of fabric, shapes and painting designs. However, the devolutionary process, so obvious in the ceramic tradition, marks the departure from the established classical tradition, which assigns it a separate status of being late phase of the Harappa. This trend is also conspicuous so far as the decoration of pottery is concerned. As a result the painting repertory is strictly restricted to linear paintings and similar incision designs. There is, of course, no dearth of such instances, however, some of them are discussed below.

This is a sherd of vase with pinkish pale and dull red surface bereft of any slip whatsoever, painted with black bands enclosing cross-hatched lines forming a ladder pattern from Bhawad-II (Fig. 5.23:8) cf Mitathal-II B[47], PLS. IX: 4; XI: 6. There is another sherd of a vase of thin section with dull pale red surface devoid of any slip and decorated with parallel horizontal lines in light black colour (Fig. 5.24:6). Similarly there is another piece of a pot of medium fabric containing only very faint and thin black lines as decoration (Fig. 5.25:9). Such linear paintings are so common in this phase to be witnessed frequently everywhere. This shows the general state of decoration technique where neither the colour applied is fast nor slip is visible nor the painting designs are rich in any manner. What we witness is a growing adoption of incision decoration technique. It seems that to make up this deficiency of the painting system they applied such incision patterns as rope marks (Fig. 5.27:5), thin horizontal parallel lines (Fig. 5.29:10-11) cf Bara[48], figure 103:6, 13; flowing thin horizontal lines (Fig. 5.29:12) cf Bara[49], figure 103:8, 15; horizontal finger marks (Fig. 5.30:3, 5-6); incised wavy lines enclosed by horizontal lines (Fig. 5.32:16); incised groups of wavy lines (Fig. 5.32:4, 15); panel of incised horizontal lines intersected/slashed by diagonal lines (Fig. 5.32:12); incised oblique lines forming chevron pattern (Fig. 5.32:7,10) etc. These marks are witnessed almost as a rule on the exterior of ceramic ware of this phase, everywhere, in the Harappan zone.

The succeeding phase produced a distinct pottery after which the period itself is designated PGW phase. The ceramic assemblage from this area underlines the extant undercurrents suggesting conspicuously different stages of the industry as could be noticed elsewhere also in this regard. And this is also visible in respect of the decoration of the pottery. The decoration process is consisted of paintings and incision designs. Of course, the paintings are executed in jet black colour over polished grey outer and inner surfaces of thin walled and well-fired ceramic ware. Black horizontal bands in singles and multiples (Figs. 5.33:1-3, 5-6, 8; 34:2-4, 10; 35:1, 6) on the rim and interiors; vertical lines in group of twos and multiples (Figs. 5.33:1-2,4; 34:2) cf Parihar[50], PL. IV-1, Hastinapur[51], figure IV: 39, 42-43; Kanpur[52] district, figure 51:15, 17, 23, Bhagwanpura[53], figure 9, Tripathi[54] figure VII: 7A, B; oblique strokes (Figs. 5.33:1; 34:5) cf Atranjikhera figure: 8, Hastinapur,[55] figure III: 20, 26; multiple dots (Figs. 5.34:3, 5; 70:17-18) cf figure VI: 5[56]; multiple circles with a dot in the centre of inner circle (Figs. 5.34:7; 70:4-5; 70:7-8,11,16) cf Hastinapur[57] figures 11:8, 15, VI: 12A, B, VII: 8; concentric and intersecting circles (Fig. 5.36:1) cf figures[58] VI: 13, 18 VII: 1A, VIII: 25; multiple circle (Figs. 5.33:3; 34:1); Loops (Fig. 5.33:4) cf figure IX[59]; panels of arcades (Fig. 5.34:8); cross-hatching forming ladder/chequered pattern (Fig. 5.34:11) cf figure[60] VII:22-23 constitute the main body of decoration repertory. Besides in certain instances of application of incision technique is also quite obvious. However, the decoration of this nature is mainly confined to very sharp incised and parallel horizontal lines (Fig. 5.38:1,11,20,24) and grey perforated pottery is available from this area (Fig. 5.31:5-7) cf Ahichchhatra[61] PL. IV-A; regular oblique strokes stemming from a horizontal deep and sharp line on thin red pottery with a round perforation of the same class (Fig. 5.31:5) forming a leaf pattern. In one instance the incision is done in an irregular manner enclosed by a panel of incised cross-hatched lines (Fig. 5.32:11). In another case the lower part of a miniature handmade vase is perforated with perhaps needle like object (Fig. 5.27:7)—its shape being of indeterminate nature one is not sure of its use and utility. The incision is done mainly on the exterior of the ceramic ware. In addition in some cases perforation was also carried out to decorate the pottery (Fig. 5.31:5-7). The plain grey ware is in abundance. The shapes and surface treatment of these potsherds (Figs. 5.39-41) shows deteriorating condition of the art of pot-making, perhaps marking the subsequent stages. Such specimens have been produced by settlements located in the Khadar belt, which seem to represent the second and third stages of the process of the spread of human activities. There are some pieces of black slipped (Figs. 5.38:13-14; 39:6,15; 40:1,11) and dark grey coloured pottery (Figs. 5.38:16, 21; 39:8-9). Likewise there are some other very interesting examples of this ware, but only two of them are demonstrated here. The first case (Figs. 5.36:1; 37:1) showed in these photographs displays the two sides (interior and exterior) of the sherds. The internal side is partly pale red and greyish with marks of intersecting circles, whereas the outer surface is pale red. Similarly the second sherd is dark grey in colour from inside while the exterior is partly pinkish red or combining pinkish red and light grey shades. These two sherds seem to hint at the lack of requisite control over the firing process and the resultant impact of pottery in terms of shade. This could also perhaps provide an insight in regard to the nature of the substance used for prefiring decoration. The colour contrast and brightness of this substance seems to have been subjected to the degree and quality of firing process. As in the present case obviously, the firing is not upto the mark and therefore, the colour of the painting and surface is also accordingly light. Whereas in normal firing conditions the usual colour is also of standardized nature. This shows the decorative substance when applied for paintings was of raw nature. To obtain its mature shade (jet black) and quality appropriate firing was called for and thus, the raw decorative substance was supposed to turn out to be a remarkable fast black colour only after proper firing for decorating the pottery of this period. Besides such potsherds being thin in section bear close resemblance to the Ganeshwar and Bara wares. A detaited and comparative study of the pottery could perhaps provide some insights into the early/formative stages of this ware. This should be treated as a proposition and not a venture to unravel the mystery of making this pottery.

During the RW-NBPW phase the ceramic industry in this section of the area does not present a very healthy situation because the pottery in general is plain and plain dull red without any painting. However, there are some examples of incision decoration, which also remained highly limited and strictly restricted to a few linear patterns such as horizontal finger marks on medium to thick red ware (Fig. 5.32:1-2), incised wavy lines enclosed by horizontal lines (Figs. 5.29:3-4,14; 32:3-5); stamped design (Fig. 5.32:6,8,9,14) cf Bijnor[62] and Chak86[63], stamped red ware sherds figure 22; double panels of verticals and chevron pattern around neck of thick ware collared jar (Fig. 5.27:9). This shows the surface treatment technique, utilitarian pottery shapes and potting techniques remained simple in general in this area.

Terracotta Objects

The terracotta objects also form a significant constituent of cultural complex. These consist of a range of objects and represent different periods. A brief analysis of such objects is attempted in what follows:

Fig. 5.43: Terracotta Discs

The discs of this nature have been discovered from Harappan, PGW and early historical settlements. These are basically round in shape and represent different denominations, chiefly designed as playthings for children. These items are discussed below:

1. Specimen of a round terracotta disc of medium size, red ware, uneven plain red surface, varying thickness of sides, dull red core. Its diameter is 4.5 centimetre and belongs to late Harappan phase from Sisana-I. If we go by its inner and outer surfaces it seems to have been carved out of a broken piece of pottery.

2. Round pottery disc made of well-fired broken potsherd. It is 4 centimetre in diameter with uneven sides and surfaces. The preparation of clay is not good and displays an admixture of extraneous material like small pieces of calcium carbonates as usually observed in late Harappan pottery, from Gangesar-I.
3. Round disc in red ware with uneven and obliterated sides, bearing faint signs of strokes along outer edge like sun pattern. It is 3.8 centimetre in diameter and seems to belong to red ware group of PGW, from Harsana Kalan-I. Such discs have been reported from Atranjikhera, Allahpur and Noh.[64]
4. Specimen of red ware disc with a continuous chain of incised strokes around outer periphery forming sun-rays pattern. It is dull looking in appearance and comes from early historical site of Bali-II. Such items seem to have constituted playing repertoire. cf. specimen from Noh, Allahpur and Attranjikhera.[65]
5. Small disc of 3 centimetre diameter in smoky shade with uneven sides and surfaces, from early historical site of Harsana Kalan-II.
6. Rectangular disc with two holes, dull red ware, 4.5 centimetre in length and 2.5 centimetre in thickness, from Baroda-III.
7. Broken hind portion of a bull in red ware and dull plain appearance. The figurine is placed in up side down position, from Ahulana-II, a late Harappan site.
8. Figurine of an ox in dull red and plain surface, a specimen of poor workmanship, perhaps a plaything for children as noticed, even in present times, from Rindhana-V, a Painted Grey Ware site.
9. Oval shaped dice, dull red ware, plain red surface, with ten holes arranged in three lines. The outer two lines are consisting of four holes each and the central one contains only two. The longer axis are 5.5 centimetre and approximate width is 3.5 centimetre. It is two centimetre in thickness at the centre, from Rindhana-I.

Fig. 5.44: Terracotta Bangles

There are a few specimen of this item, which are mainly collected from early Harappan settlements. These are displayed mainly in this figure as follows:

1. Fragment of a bangle round in section, plain red ware and surface, greyish core, devoid of any decoration. Its thickness is 8 mm, from Rindhana-I.
2. Small piece of a bangle, roundish in shape and section, greyish core with flattish sides in plain red ware. Its size is 2 centimetre in length and 7 mm in thickness, from Ghadwal-I.
3. Broken piece of a bangle in pale red ware, dull core and plain surface, and roundish in section. Its dimension runs into 6.1 centimetre, which speaks of its sturdy nature, from Rindhana-III.
4. A bangle piece 4 centimetre in length and 1.2 centimetre in thickness, round shape and section, light red outer surface, grey core, a sturdy and strong specimen in comparison to previous examples, from Baroda-III.
5. Fragment of a bangle with 2.5 centimetre in length and 0.5 centimetre in thickness, plain red ware, greyish section. The outer top portion is pinched up to form cogged design, from Samri, cf Mitathal II B, PL. XXV: 1, 2, 4, 6, 8.[56]
6. Broken bangle piece 4 centimetre in length and 0.6 centimetre thick, red surface, grey core, treated with light red wash, flattish sides and roundish top, from Nurankhera-I.

Figs. 5.45-47: Terracotta Cakes

Apart from the aforesaid objects some terracotta cakes have been picked up from various sites. These represent the various stages of the Harappa civilization with the change in their size, shape and quality of firing. In the first phase–early Harappan Mitathal-I these cakes are either elliptical or circular with tapering sides, smaller in size and thicker in the centre (Figs. 5.45:1,5; 46:2, 4). Besides some of them also bear finger tip depressions on both the faces (Fig. 5.46:1). And still there is one more example from this period with elliptical shape in dull grey ware. This is very interesting, since this bears an incised impression of an uneven toed animal hoof, probably of a cow (Fig. 5.45:7). This comes from Rindhana-III, an early Harappan mound. In the succeeding mature Harappan phase the art of making these cakes experiences transformation as their size is enlarged and the section is accordingly flattened making their ends/sides relatively round and sharpened (Figs. 5.45:2, 4; 46:3; 47:1, 3). While some other discs have also been gathered in the process of exploration from Gangesar-I and Ahulana-I. These are triangular in shape with rounded sides (Figs. 5.46:5-6; 72:5) signifying changes taking place in the ceramic industry. This marks the evolutionary process.[67] However this trend was reversed in the subsequent late Harappan phase. Because the discs belonging to this phase again underwent a change to adopt dull appearance with uneven and irregular sides, surface and shade (Figs. 5.45:3; 47:2). Thus, these present beginning of a devolutionary process. Such discs have been reported from Mitathal-II B[68], Bargaon and Hulas[69] figure 115. Moreover these are invariably mixed with chaff as degraissant all through. Similar observation was made by Bhan in this regard.[70] Hence, with the changing fortunes this art also experienced corresponding alterations.

Figs. 5.48-49: Terracotta Beads

Several terracotta beads were also recorded from various settlements belonging to different stages of protohistoric and early historical periods. To have an idea of the art of bead making and their uses a brief illustration of some of the selected and representative specimens is attempted in the following paragraphs.

1. This is a standard bicone circular bead in red ware with pin-hole and signs of fugitive red polish, from Ghadwal-I (Fig. 5.48:1).
2. This is a bicone circular bead in dull and plain red ware, from Ahulana-II (Fig. 5.48: 2) cf. Mitathal[71], PL. XXIII: 7.
3. This is a slightly enlarged bicone circular bead in plain dull red ware and decorated with an engraved star sign/ grafiti mark on the outer surface near the hole, from Bhainswan (Fig. 5.48: 3).
4. This is a short bicone circular small bead in dull red ware from Kathura-I (Fig. 5.48: 4).

5. This is a piece of a conical and truncated beads in grey ware from Nurankhera-II (Fig. 5.48: 5).
6. This is a short bicone circular bead with truncated faces in plain dull red ware from Rindhana-I (Fig. 5.48: 6).
7. This is a bicone circular bead with truncated faces in grey ware from Ghadwal-III (Fig. 5.48: 7).
8. This is a short bicone circular bead with truncated faces in burnt grey ware from Kakana (Fig. 5.48: 8) cf. Mitathal[72] PL. XXIII: 5.
9. This is a short bicone circular bead with truncated faces in plain dull red ware from Kathura-I (Fig. 5.48: 9) cf. Dadheri[73] figure 30.4.7 and Mitathal[74] PL. XXIII:3.
10. This is a specimen of a ghat-shaped bead in dull red ware (Fig. 5.48: 10). This being a landmark of the PGW phase has been produced from several sites such as Khanpur Kalan-I, Gumad, etc. Fig. 5.48: 10). cf. Dadheri[75], figure 34, 4: 6, Hastinapur[76], figure I (unstratified). Similar beads have also been reported from Atranjikhera, Ahichchhatra, Allahpur, Noh and Rupar[77].
11. This is a short bicone circular and internally bevelled wide holed bead in dull red ware from Ghadwal-I (Fig. 5.48: 11).
12. This is a broken conical bead in grey ware from Ghadwal-III (Fig. 5.48: 12). Such beads belong to PGW levels.
13. This is also a partly broken short biconical truncated bead in red ware with grey core from Busana Kalan (Fig. 5.48: 14).
14. This is a conical ant truncated bead in grey colour from Panchi Gujjaran (Fig. 5.48: 15).
15. This is a conical and truncated bead in grey ware with an incised line around the hole from Sisana-II (Fig. 5.49: 1).
16. This is a fragmentary biconical and truncated bead in grey ware decorated with incised ring around the hole from Rindhana-V, a PGW site (Fig. 5.49: 2).
17. This is another fragmentary biconical and truncated bead

in grey ware decorated with three black and one incised line on the bevelled shoulders, from Kakana (Fig. 5.49: 3).

18. This is a conical bead in plain red ware from Jawahara-I (Fig. 5.49: 8).
19. This is a standard conical and truncated bead in grey ware from Gangesar-II (Fig. 5.49: 9) and Jawahari-I. This type has been fond from PGW sites[78].
20. These are terracotta small balls in red ware from various sites and might have been used as playthings in the past (Fig. 5.49: 4-6, 10). Similar balls have been reported from other PGW sites also[79].
21. This is another terracotta, red polished ball with five engraved lines dividing its body in equal vertical compartments from Ghadwal-I (Fig. 5.49: 7). This bears similarity with cotton seed capsule, musk-melon and seed of anwala fruit.
22. This is a broken long barrel circular pin-holed bead of red ware from Ghadwal-I (Fig. 5.49: 11).
23. This is a small oval-shaped terracotta ball, polished in red, well-fired and bearing five engraved lines dividing the body into five compartments, which give multiple impression of its being either a musk-melon or cotton-seed capsule (Fig. 5.52:8). And this could well have been used as a plaything by children, from Ghadwal-I.

Fig. 5.50: Terracotta Wheels

Like other terracotta objects a few toy-cart wheels were also gathered from various sites associated with different cultural phases. Among them mention could be made of some mature Harappan wheels, which, of course, demonstrate an advanced stage, a characteristic feature of this phase. The wheels of this phase could easily be identified with a prominent central hub as witnessed usually everywhere in this phase. Besides these are well-fired and decorated with black bands over red slip (Fig. 5.48:1) from Baroda-III. Similar instances have been reported from Bargaon[80]. In addition, one may also state two more such specimens from Ghadwal-I and Rindhana-I. These bear raised central hub (Fig. 5.48:2-3). Such examples have also been quoted from Bhagwanpura-I and Hulas.[81] But in the late Harappan phase the precision of the yester-years is conspicuous by its absence if shape, fabric and appearance of such objects are any indicator. In this phase these wheels become smaller in size, thicker in section and dull in appearance. The tradition of central hub is not visible. This seems to have already been replaced by raised portion around the hole as noticed above. In some cases even broken pot sherds have been transformed into wheels (Figs. 5.48:4; 31:4). Similarly these are devoid of any painting and incision decoration. This shift is in consonance with changing overall cultural environment.

Fig. 5.51: Stone Objects

Some objects made of stone were also recovered from surface of different settlements during course of exploration. Most of these objects are made of sand stone and constitute the household repetoire. Some of the items are discussed below:

1. This is a small round ball of sandstone, which bears signs of pecking technique as indicated by uneven surface. It bears a brown band around. Comes from Busana Kalan, a late Harappan settlement. This could have been used either for playing or as weight.
2. Fragment of a medium-sized ball of red sandstone with relatively smooth surface. This was collected from Gangesar-I and might have been put to similar uses as mentioned earlier.
3. This is an oval shaped/roundish disc in red sandstone with uneven surface. Its length is 6.5 centimetre and width is 65 centimetre from centre. Its thickness varies from one point to other which makes it uneven and dull looking, from Jawahara-I.
4. This is an elliptical ball of red sandstone with flat and smooth base, perhaps used for grinding as is witnessed these days, from late Harappan site of Banwasa.

5. This is a specimen of elliptical shaped pestle of red sandstone with roundish ends and uneven surface, from Jawahari-I, a PGW site.
6. This is a rectangular pestle of red sandstone with 8 centimetre in length and 5 centimetre in width, bearing thumb impression on one side, while the other side is broken. Its both ends are roundish in shape with uneven surface suggestive of its use for grinding raw substances, from Bhatgaon-I.
7. This is a fine specimen of a rectangular/wedge-shaped pestle in black slate stone bearing grinding signs on both the roundish ends. The attractive smooth surface and shape do speak of good workmanship. Likewise smooth roundish ends remind of its specific grinding sophisticated functions, from Ghadwal-II.
8. This is a cubical weight in red sandstone with 3 centimetre long sides, a lone specimen of this kind. Comes from Baroda-III. Usually in the mature Harappan phase, especially at advanced centres, such cubical weights are very common, but generally found to be made of fine grained chert stone. Such weights of red sandstone are witnessed replacing the chert weights either in peripheral mature Harappan sites or in succeeding late Harappan levels. And are considered either a sign of low connectivity with the advanced mature Harappan centres or a mark of a devolutionary process trigered off in succeeding late Harappan levels. But in this connection it is desirable to take note of the fact that the Harappan settlements in this area are primarily village sites well away from urban centres. And at the same time these rural settlements appear to have in some respect contemporary to urban Harappan phase. Hence, these were bound to have some direct or indirect feel of the urban tradition, but at the same time it was not necessary for all these settlements to have equal level of connectivity. And the cubical weight of sandstone is a testimony to suggest such a variation in the number, quality and raw material of these things. And such differences are not uncommon even these days between the hinterland villages and advanced metropolitan

cities despite the fact of their being contemporaries. Hence, if this is viewed in this light the currency of cubical weights of sandstone at some settlements is indicative of their rural nature and thereby the product of the uneven level of development. This fact has further been born eout by the general composition of artifacts and their raw material as discussed in the previous paragraphs. As all the objects are made of locally available raw material in the form sandstone, this in turn underlines the absence of any effective mechanism for long distance trade and thereby the lack of well geared interaction to provide for uniform process of development and dissemination of quality products over the vast area alike. This could well have been coupled with non-affording economic condition of the people in general in such rural interiors. The limited circulation, simple nature of these objects and their local raw material point further towards a situation where people seem to have remained conversant with the technique of their manufacture and use but only with the raw material of local origin. With the result the people had to contend with what was locally available irrespective of their quality and raw material. This is indicative of the continuing adaptive attitude and the utilitarian approach of the people towards life on the whole.

Semi-Precious Stone Beads

A few beads made of semi-precious stones were also picked up from different settlements in course of exploration. These beads are very attractive in their shape, appearance and technique. These, however, present a limited variety in shapes and raw material, but surely signal not only the currency of these beads and knowledge of the semi-precious stones, but also the mechanism, to obtain them and likings of the people for them. A description of these beads with their double view is as follows:

Figs. 5.52-53

1. This is a long barrel elliptical bead of agate in dark brown colour with inbuilt strip decoration in different colours and provided with a hole to insert a thread through it. It appears to have been a specimen of fine workmanship as the outer surface is very smooth, glistening and shining. Comes

from Ghadwal-I. Similar beads have also been recorded from unstratified levels at Mitathal[82], PL. XXII-1.

2. This is a long barrel circular bead of toggle shape made of carnelian in light brown shade, pierced vertically in the middle to insert a thread. Its outer surface is best finished to give a shining look and speaks of highly established art of manufacturing. Comes from Bhawar-I. It is comparable in shape to some beads from Mitathal[83], figure 15:5-7.
3. This is a long barrel circular bead of agate and decorated with inbuilt circular bands in black as well as in light brown separated by white circular thin bands. It is pierced through horizontally to insert a thread. It gives a variety of shades at different angles. It is also a product of fine technique and workmanship, from Ghadwal-III. Similar bead has been reported from Mitathal-IIB, figure 15.8[84].
4. This is a medium sized barrel circular bead of agate from Rindhana-I, decorated with alternate bands and lines in white, dark-brown, light-brown and black with a thin horizontal hole. Similar unstratified bead has been reported from Mitathal[85], figure 14:15.
5. This is a long barrel lenticular bead of agate from Baroda-III, a hole pierced through horizontally, decorated with alternate following grey, brown and white thin bands forming an interesting ripple design on the outer surface of the bead. In respect of look, shape and technique it is of remarkable nature. It is comparable to those gathered from unstratified levels at Mitathal[86], figure 15:2-3.
6. This is a small barrel circular bead with a narrow horizontal hole, decorated with white and brown registers/bands. These give different impressions and form designs at different angles. The outer surface is very smooth, from Ahulana-I cf Mitathal[87], figure 15:15.
7. This is a specimen of cylindrical circular bead of agate from Gangesar-I decorated with regular white circles over black surface forming an attractive design. The surface of the bead is divided into two halves consisting of four circles each.

Figs. 5.60 and 69

1. This is a parallel sided blade of Chert from Ghadwal-I.
2. This is a lower part of a broken flake-like chert blade from Rindhana-I.
3. Several micro-beads of steatite were collected from Ghadwal-I (Fig. 5.69: 4-10).
4. This is a cogged wheel/bead of Chert with a central hole from Ghadwal-I.
6. This is a pendent like circular bead of chert in light cream colour, from Ghadwal-II.
7. This is circular bead of carnelian in dark brown colour, from Ghadwal-I.
8. This is a specimen of long barrel circular carnelian bead in dark brown colour, from Rindhana-I.
9. Micro-beads of carnelian alongwith a piece of carnelian have been gathered from Ghadwal-VI (Fig. 5.69:1-3).
10. This is a fragment of a steatite bead, barrel circular in section, dark grey in colour, having a very smooth outer surface and devoid of painting, from Rindhana-II.

Faience Objects

This substance seems to have remained very favourable raw material for the Harappans, especially the late Harappans, to fashion out a variety of items according to their requirements. In the course of exploration a fairly large number of faience objects was gathered to add to the cultural assemblage. But surely among them the predominent position is enjoyed by broken pieces of bangles. A detailed description of some of the illustrated objects is as follows:

Fig. 5.61: Faience Objects

1. This is a fragment of a faience bangle with raised external medial rib, triangular in section, bluish in colour, decorated with incised oblique lines on either side of the rib forming a chevron pattern and single horizontal parallel lines along the edges, from late Harappan settlement of Sargathal.
2. A strap like board bangle piece of externally flattened section, greenish colour, decorated with criss-cross incised

lines forming irregular chequered pattern, from late Harappan settlement of Sargthal.

3. A piece of a thick and broad bangle, oblate in section, sky-blue in colour, decorated with oblique incised irregular lines, from Ghadwal-III.
4. Fragment of a bangle, oblate in section, greenish in colour, decorated with deep incised evenly spaced oblique lines, from Sisana-I and Ghadwal-III.
5. Piece of a strap like broad bangle, oblate in section, greenish shade, decorated with a panel of irregular incised lines between two horizontal parallel incised lines on both sides, from Rindhana-III.
6. Piece of a bangle with flattened section, bluish in colour, decorated with cross-hatched lines forming a fine chequered pattern giving the impression of knobbed design, from Baroda-II cf Mitathal,[88] PL. XXIV: 16.
7. Fragment of a bangle, flattened oblate section, light blue in colour, decorated externally with oblique incised lines emerging from a horizontal line, from Sisana-II.
8. Fragment of a bangle, greenish in colour, flattened oblate section, decorated externally with oblique lines, from Ghadwal-III.
9. Fragment of a broad strap like bangle, sky-blue in colour, flattened oblate section, decorated with oblique incised lines enclosed by two horizontal lines, one on each side, from Ghadwal-III.

Fig. 5.62

1. Standard bicone circular bead with medial rib in greenish tinge, from Ghadwal-II, cf. Mitathal[89], PL. XXII:17.
2. This is a long barrel circular bead in sky blue colour from Rindhana-III, cf Mitathal-II B, PL. XXII: 10 & 11[90].
3. This is a bicone circular bead of faience in sky-blue colour, from Banwasa (Fig. 5.60: 9).
4. This is a broken barrel circular bead in dull greenish colour, from Sisana-II (Fig. 5.60: 10).

Fig. 5.63

1. Fragment of a bangle, sky-blue in colour, thin circular in

section, devoid of any decoration from Ghadwal-III. Some other settlements like Ahulana-II and III, Gangesar-II and Gohana-I have also produced such bangle pieces. These vary only in their thickness and size.

2. Fragment of a heart-shaped bangle, thin circular in section, greenish in colour and glazed white, from late Harappan settlement of Nuran Khera-II cf Mitathal-II B,[91] PL. XXIV:17.
3. A piece of a bangle, brownish in colour, thin triangular in section, decorated with incised oblique lines on the top, from Ghadwal-II.
4. Fragment of a bangle, sky-blue in colour, thin triangular in section with rising central point assuming shape of a medial rib, devoid of any decoration, from Baroda-II. Such evidences have also been gathered from other settlements like Nuran Khera-II, Sisana-II, Pipli-I and Chatia Deva.
5. A small piece of a bangle, sky-blue in colour, planoconvex section, decorated with incised oblique lines enclosed by horizontal lines, glazed blue, from late Harappan settlement of Dodwa. Such bangle pieces have been collected from Bhawad-II, Ghadwal-III, Bichpari-I and Banwasa also.
6. Fragment of a bangle, blue in colour, medium circular section, decorated externally with a series of incised vertical strokes enclosed by two horizontal lines forming a ladder pattern, from Bohla.
7. Fragment of an indeterminate object, appearing like either a leaf or fish filled with oblique strokes. It is in sky-blue colour. Comes from Ghadwal-III. This is the sole example of its kind.
8. A broken piece of a bangle, sky-blue in colour, rectangular in section, devoid of any decoration, from Ghadwal-II and Bhawar-I.
9. Fragment of a rim, possibly either of a miniature cup or a bowl with externally sharp bevelled vertical rim, decorated externally with round incised lines, from Ghadwal-II.
10. This is another broken sharp edged rim and devoid of any decoration whatsoever, from Ghadwal-III.

Fig. 5.64

1. Fragment of a bangle, ridged externally in the centre, triangular in section, with bluish tinge, decorated externally with three parallel horizontal incised lines on either side of the ridge with regular notches on top, from Chatia Deva.
2. Fragment of a bangle with externally ridged centre, triangular in section, greenish in shade, decorated with incised oblique lines on one side with notches at regular intervals on the pinched up ridge, from Ghadwal-III.
3. Fragment of a bangle with externally high central ridge, triangular in section, sky blue glazed green, decorated on the exterior with a set of two incised parallel horizontal lines, from Baroda-III.
4. Fragment of a bangle, sky-blue in colour, round in section, decorated with faint oblique lines enclosed by single horizontal line on each side, from Ghadwal-III.
5. Piece of a bangle, circular in section, sky-blue in colour, decorated with oblique lines on the exterior, from Rindhana IV, a late Harappan site, Cf. Mitathal-IIB,[92] PL. XXIV:1.
6. Piece of a bangle, glazed white, externally corrugated with four lines, from Ghadwal-III cf Mitathal,[93] PL. XXIV:14.
7. Fragment of a bangle, blue in colour, oval circular in section with almost vertical sides, decorated with oblique cut marks on the top, from Bhawar-I.
8. Fragment of a bangle, sky-blue in colour, planoconvex in section, decorated with a panel of incised vertical strokes enclosed by a group of horizontal parallel lines, glazed blue, from Sargthal and Ghadwal-III.
9. Fragment of a bangle, planoconvex in section, blue in colour, decorated on the exterior with alternate blue and white glazed parallel lines, from Baroda-III.
10. Fragment of a heart-shaped bangle, planoconvex section, greenish in colour, decorated externally with incised round lines forming grooved surface, from Baroda-III.

11. A small piece of bangle, blue in colour, thin round section, decorated externally with opposite oblique strokes forming fine Chevron design, from Pipli-I.
12. Piece of a heart-shaped bangle, green in colour, decorated externally with oblique strokes forming a chevron pattern. From Bhawar-II. Similar specimens have also been recorded from Ghadwal-I, II and III, Rindhana I and IV and Bichpari-I.

The foregoing description of varied faience objects is very enlightening on various counts. First of all it is indicative of an overwhelming preference for this substance, which forms one of the main raw materials to produce finished goods. This also implies that the other raw materials of better availability, common in Harappan Urban Centres, seems to have been substituted by this substance in this zone either because of non-availability of the other substances or owing to interrupted and ineffective trade mechanism or due to non-affording economic condition of the rural people to go for the luxury goods. The growing use of this material for a varied and rich variety of items makes it amply clear that not only people had likings for decoration but also they were resourceful in terms of techniques and alternatives to satisfy their tastes. Hence, in absence of fine quality raw materials they preferred slightly inferior matter to produce fine looking finished goods to serve their needs. Thus, according to the changed environment people adapted themselves to negotiate successfully in the given situation in order to make their life still pleasant.

Copper Objects

During exploration some copper objects were also collected from different settlements. Their range is limited and mainly confined to the Harappan settlements. Almost all of them ware corroded badly. In a few instances corrosion marks are still visible. A brief description of these objects is as follows:

Fig. 5.65

1. This is a broken red antimony rod like object of copper. Its upper end is slightly sharp and the lower portion is blunt and broken. This could well be a part of a distaff also. The outer surface is uneven, from Harsana Kalan-I.

2. This is another piece of a copper antimony rod like object. Its one end is quite smooth and the other is a bit sharp, from Gangesar-II.
3. This is a fragment of a copper bangle in circular section with pinched up formation on the top side just like that of the heart-shaped bangles of faience, from Bohla.
5. This is a tiny round ball in copper from Chhapada. Its use is unknown.
6. This is broken piece of a pointed pinched up arrow head like copper object from Busana Kalan.
7. This is a broken piece of a flat copper bangle with slightly bulging formations on top at regular intervals from Ahulana-III.
8. This is a bicone circular bead of copper. Its outer surface is badly corroded. From Sisana-II (Fig. 5.64:5).
9. This is a broken triangular copper arrow head of thin section and smooth surfaces. From Kakana.

Figs. 5.67-68

1. Copper Kushan coin circular in shape, 2 centimetre in width, 0.4 centimetre in thickness and weighs 6.82 gram.

   Obverse symbol: Standing Kushan king and holding something in his right hand.

   Reverse mark: An indeterminate figure with a chain of regular dots around the edges, from Harsana Kalan-III.

   Remark: It resembles the later Kushan coins from Sonipat as reported by Phogat.[94]

2. Copper Kushan coin circular in shape, 2 centimetre in width, 0.4 centimetre in thickness and weighs 7.33 gram.

   Obverse symbol: Standing Kushan king.

   Reverse mark: Standing Lord Shiva before Nandi (sacred bull of Lord Shiva), from Harsana Kalan-III.

   Remark: It resembles Later Kushan coins from Sonipat[95]

3. Copper Kushan coin circular in shape, 2 centimetre in width, 0.4 centimetre in thickness and weighs 7.2 gram.

Obverse symbol: Standing Kushan king offering on altar with his right hand.

Reverse mark: Standing human figure on a platform from Harsana Kalan-I.

4. Copper Kushan coin circular in shape, 2 centimetre in width, 0.4 centimetre thickness and weighs 6.58 gram.

   Remark: It is like the first coins.

5. Copper coin of semicircular shape, 1.4 centimetre in width, 0.3 centimetre thickness and weighs 3.45 gram.

   Obverse symbol: There is an obscure symbol like hill.

   Reverse mark: Indeterminate mark, from Bali-II.

6. Copper piece with 1.3 centimetre width, 1.7 centimetre length and 0.5 centimetre thick and convex sides. From Datauli. Its weight is 10.57 gram.

7. Broken small piece of copper, perhaps a coin, of 'U' shape with 1.1 centimetre in length, 0.9 centimetre in width, 0.2 centimetre in thickness and weighs 0.450 gram, from Harsana Kalan-I.

8. Copper/iron bent bar with pinched ends with 2.7 centimetre in length, 0.5 centimetre thickness on one end and 0.3 centimetre on the other, 1 centimetre in width and weighs 6.40 gram, from Harsana Kalan-I.

9. Copper circular coin with swollen and uneven surface, 2 centimetre in width, 0.3 centimetre in thickness and weighs 10.73 gram. An obscure symbol on the reverse, from Mandori.

Miscellaneous Objects

1. This is an antimony rod or a bone point in dark brown colour and with very smooth surface. Its lower side is broken, from Harsana Kalan-I (Fig. 5.65: 4).

2. This is a tubular shaped broken piece of some substance looking like iron, but not in fact. Its shape appear like a long rivet used for joining wooden parts of some object, from Akbarpur Barota (Fig. 5.65: 8).

3. This is a heavy round ball of iron with slightly uneven surface, from Jawahari-I (Fig. 5.65: 10).
4. This a circular disc of some substance like soft stone in cream colour, from Ghadwal-II. This might have been used as a plaything (Fig. 5.66: 1).
5. This is a broken terracotta Yaudheya coin mould from Harsana Kalan-I (Fig. 5.66:2).
6. This is a piece of iron ore from Harsana Kalan-I (Fig. 5.66: 3).
7. This is a broken base of a dish-on-a stand from Ghadwal-I (Fig. 5.72:4)

Conclusions

The foregoing description provides us a comprehensive picture of the material culture in various phases of period under discussion. This period, of course, begins with the inception of human life sometime in the second half of the so-well-known first phase—the early Harappan of the Indus Civilization. Since this forms the initial stage of the beginning of human life the nature of the material culture is in consonance with that. The material remains are, as we have discussed above, consisting of pottery and some other terracotta objects. Among them naturally, pottery assumes a dominant position for its greater frequency. In this context one marks a rudimentary stage of the art of pot-making wherein maturity, perfection and standardization, as a natural corrollary, are visibly dificent. The pottery is simple, both handmade and wheel-turned, usually lacking in surface treatment, painting designs, evolved pottery shapes, uniformity of firing process (resulting in ceramic wares of different colours) and universality of these features in terms of the greater part of the Indus Valley region. In addition to this circulation of some terracotta beads and bangles to enrich the cultural repertoire is to be met with. However, these objects are also equally very simple and limited, especially in absence of any digging, in number. This points among many things to the localized nature of this culture. On the whole the material millieu is not only simple, but also strictly restricted to utilitarian nature and approach.

As soon as one embarks upon the succeeding mature Harappan phase one awaits to witness all round evolutionary process. This trend is quite discernible in every single and collective respect as well. This is, however, noteworthy that even in this phase the main components of the material culture and their respective frequency seem to have not undergone any comprehensive transformation to undermine the ongoing overwhelming predominant position of pottery among the cultural artifacts.

The pottery of this phase is invariably wheel-thrown of bichrome group (black-on-red), unlike the polychrome ware of previous phase, which appears to be superior to polychrome group in respect of potting, firing, surface treatment, painting designs, pottery shapes and near uniformity in adoption of these points in the region as a whole. This signifies the standardized position attained by the art of pot-making in this phase. This would further be ascertained by a mere look on the pottery. A similar trend of transformation is also observed in other areas of cultural activities. Terracotta beads and bangles of the yester-years seem to have been growingly replaced with those of chert, carnelian and agate, all fine quality stones. This marked difference is not only visible in the quality of raw material, but also in the technique employed to make these items with precision. So far as the frequency of these objects is concerned, it is, of course, considerably restricted, yet, it marks departure from the previous phase. The low frequency of these objects could perhaps be attributed to the poor stimuli, among other things, than the major centres for development and interaction to share new ideas, techniques and products. We also do not known whether these objects were produced locally or received through hinterland trade. Likewise in this very phase the currency of cubical weights and pestles, irrespective of their raw material, point in the same direction. The availability of terracotta figurines and dice with holes do speak of their sense and sources of recreation. The evolutionary process is further highlighted by the relatively large number and size of settlements and cultural depositions suggesting relative stability. It seems, therefore, that the cultural process began sometime in early-Harappan levels seems to have relished some taste of the development in this phase.

Coming to the late Harappan phase, the last phase related with the Indus Civilization, the earlier position seems to have been reversed. In this phase a devolutionary trend is quite obvious in all respects, except the ever commanding position of ceramic ware among the cultural components. The precision of the yester-years, however, is not visible in regard to the ceramic industry because the clay is not well-levigated as presence of foreign elements is quite obvious, firing process is not uniform giving different colours to their sections in absence of precise control over the firing process, surface treatment is not good as red polish of uniform nature is not invariably applied (which is of fugitive nature), fine shapes of the mature phase have been replaced with rough and dull shapes and finally painting designs are increasingly linear and repetitive. Incision is being adopted growingly for decoration. Similarly even the limited use of precious stones seem to have been substituted by relatively raw and inferior substances, like faience and sandstones, of local origins. The quality and frequency of playthings have also underwent drastic changes, since these amount to be handiwork of poor workmanship. Terracotta objects such as dice, toy-cart wheels with central hub and fine triangular sling balls are missing. Therefore, the hallmark—standardization, a fair uniformity and stability–of the yester-years is undermined substantially. Moreover the small size, higher frequency and relatively impermanent nature of many settlements of this phase altogether hint an increasing resurgence of regionalism and instability. This also points to the absence of an effective mechanism for interaction and coordination.

The late Harappan phase is followed by the appearance of Painted Grey Ware ceramic tradition. And the cultural complex of this period seems to have been a little bit different than what we have witnessed till now. Because after a substantial time gap a novel experiment is to be witnessed in context of ceramic industry, which is popularly known as PGW after which the period and its cultural complex as well have been designated despite the fact of its restricted circulation. The ceramic industry, nevertheless, betrays sophistication on all counts whether be it preparation of clay, thickness and colour of pottery shapes, firing technique, surface treatment, application of jet black fast colour for executing geometrical as well as natural/floral designs, etc. In fact this alone would have been suffice to indicate a level of cultural advancement and sophistication, but its highly restricted frequency, possibly for some specific purpose, is enigmatic. The major chunk of the ceramic complex is consisting of plain red ware. Perhaps designed for daytoday use, where one can easily underline fast disappearance of remanent features of Harappan ceramic industry. Therefore, this ware growingly wore a simple appearance to serve daily needs. Thus, the utilitarian orientation is on the fore. So far as other artifacts of the cultural repertoire are concerned, mention could easily be made of *ghat*-shaped terracotta beads, a salient feature of this period, round terracotta as well as broken pottery discs having incised strokes along the outer edge in a radial pattern forming, especially on the former, sun design. These items could have been used by children for playing. Therefore, it seems that barring PGW in this phase there was anything to assign it a superior position. This has again been vindicated by fair frequency of usually small settlements with flimsy depositional remains. This reminds us of the limited operational potential of the people against the given natural environment like that of the preceding late Harappan phase.

In the early historical period, which is in succession to the preceding Painted Grey Ware phase, the cultural scenario is also dominated, as usual, by ceramic industry for its higher frequency among cultural constituents. Here the ceramic industry is consisting of Red Ware and the Northern Black Polished Ware—a novel variety of deluxe pottery. Undoubtedly the circulation of the latter is very meagre, limited to only a few potsherds and confined to only a few settlements along the Western Margins of the river Yamuna and in the Bhangar. This ware could not find common currency among the people for a variety of reasons. As a result the ceramic ware for common use remained the Red Ware like previous phase. It is recalled in this context that this is rich in shapes, but surely lacking evolved variety of Harappan period. Likewise, it is invariably missing the very established norms of clay preparation, firing technique, finishing of the surface and paintings for decorations altogether. The emerging picture does not appear very encouraging so far as the ceramic industry is concerned. This, however, reiterates utilitarian outlook of the people because of, perhaps, their over preoccupation with subsistence activities. This is further explained by highly limited frequency and variety of terracotta beads and figurines as only a few terracotta conical beads and gamesmen are recorded, perhaps in absence of excavations. In addition a few terracotta discs carved out of broken potsherds are also met with to sound a similar note. These objects, in fact, despite their poor number provide an insight not only into their sources of entertainment, but also of their inferior quality, raw material and innovation. The higher number of settlements with greater dimension and thickness of cultural disposition may also indicate in the same direction. Because this explains their preoccupation with the pursuit of feeding the additional mouths by way of raising stock and food crops at a higher rate in the given vicissitudes of nature. This objective was apparently not very easy to realize, especially in those days. And naturally this would have demanded more of human labour, time and hence their preoccupation with such activities. This was obviously bound to affect their very creative orientation and activities in an adverse manner. And thus, their main cultural objective seems to have hinged around the security of subsistence.

NOTES AND REFERENCES

1. Lal, B.B. and B.K. Thapar: 'Excavation at Kalibangan: New Light on the Indus Civilization', *CF*, 9, 34, 1967, pp. 78-88.
2. Bhan, S., *Excavation at Mitathal (1968) and Other Explorations in the Sutlej-Yamuna Divide*, Kurukshetra University, Kurukshetra, 1975, pp. 95-109.
3. Bisht, R.S. Banawali: A New Harappan Site in Haryana *ME*, II, Delhi: 86-88; Idem: 'Excavation at Banawali 1974-77', Possehl, G.L. ed. *Harappa Civilization*. New Delhi, 1982, 118.
4. Bhan, 1975.
5. Siwach, D.S., 'Post-Harappan Kilns (Bhattis) from Balu (Haryana)', *PIHC*, Dharwad, 1988, 45-48.
6. Bhan, 1975.
7. Bisht, 1982, *op. cit*:, pp. 113-24.
8. Bhan, 1975.
9. Singh, U.V. and Suraj Bhan 1982: 124-26.
10. Bhan 1975, *op. cit*., p. 20.
11. Sharma, Y.D., 'The pre-Harappans in Punjab', *Puratattva* 11, New Delhi 1979-80, pp. 34-35.
12. Nath, Shankar 'Pre-Harappan Potery from District Saharanpur, U.P.' *Puratattva* 11, New Delhi, 1979-80, p. 123.
13. Sharma, R.P.—'Khatoli—A Protohistoric Site in Haryana'? *Puratattva* 11, New Delhi, 1979-80, p. 124.
14. Dikshit, K.N., 'The Sothi Complex: Old Records and Fresh

Observation', in Lal, B.B. and S.P. Gupta (eds.) *Frontiers of Indus Civilization*, Delhi, 1984, p. 534.

15. *Ibid.*, p. 535.
16. *Ibid.*, p. 536.
17. Dikshit, K.N., 'Late-Harappan in Northern India', in Lal, B.B., and S.P. Gupta (eds.) *Frontiers of Indus Civilization*, Delhi, 1984, p. 263.
18. *Ibid.*, p. 262; Idem 'The Excavation at Hulas and further Exploration of the Upper Ganga Yamuna Doab', *ME* V, Pune 1981, p. 74.
19. *Ibid.*, p. 260.
20. Bhan, 1975, *op. cit.*, p. 106.
21. Sharma, 1979-80.
22. Bhan, 1975, *op. cit.*
23. Dikshit, 1984, *op. cit.*, 543.
24. Bhan, 1975, *op. cit.*, p. 106.
25. *Ibid.*, pp. 96-101.
26. Dikshit, 1984, *op. cit.*
27. Agrawal, R.C., 'Aravalli, the Major Source of Copper for the Indus and Indus-related Cultures' in Lal, B.B. and S.P. Gupta (eds.), *Frontiers of Indus Civilization*, Delhi 1984, pp. 157-62.
28. Bhan, 1975, *op. cit.*
29. *Ibid.*
30. *Ibid.*
31. *Ibid.*, p. 94.
32. *Ibid.*, pp. 31-32.
33. *Ibid.*
34. *Ibid.*
35. *Ibid.*
36. *Ibid.*
37. Dikshit, 1984, *op. cit.*, pp. 258-69.
38. Bhan, 1975, *op. cit.*, p. 36.
39. *Ibid.*, p. 46.
40. Dikshit, 1984, *op. cit.*, p. 543.
41. *Ibid.*

42. *Ibid.*
43. *Ibid.*
44. Bhan, 1975, *op. cit.*
45. Bala, M., 'Maltese Cross on Protohistoric Pottery', *Puratattva* 11, New Delhi, 1979-80, pp. 26-32.
46. Bhan, 1975, *op. cit.*
47. *Ibid.*
48. Dikshit, 1984; *op. cit.*
49. *Ibid.*
50. Lal and Dikshit, 1980, *op. cit.*
51. Tripathi, V., *Painted Grey Ware—An Iron Age Culture*, Delhi 1976.
52. Lal, M. *Settlement History and Rise of Civilization in Ganga-Yamuna Doab*, Publishing Corporation, Delhi, 1984, pp. 295-97.
53. Asthana, S. and R.S. Bisht, 'Banawali and some other Recently Excavated Harappan Sites', in Taddei, M. *South-Asian Archaeology*, Naples, 1979, pp. 223-40.
54. Tripathi, *op. cit.*
55. *Ibid.*
56. *Ibid.*
57. *Ibid.*
58. *Ibid.*
59. *Ibid.*
60. *Ibid.*
61. *Ibid.*
62. Mughal, M.R., 'The Post-Harappan Phase in Bahawalpur District, Pakistan' in Lal, B.B. and S. P. Gupta (eds.) *Frontiers of Indus Civilization;* Delhi, 1984, pp. 499-503.
63. *Ibid.*
64. Tripathi, *op. cit.*
65. *Ibid.*
66. Bhan, 1975, *op. cit.*
67. *Ibid.*
68. *Ibid.*

69. Dikshit, 1984, *op. cit.*
70. Bhan, 1975, *op. cit.*
71. *Ibid.*
72. *Ibid.*
73. Dikshit, 1984, *op. cit.*
74. Bhan, 1975, *op. cit.*
75. Dikshit, 1984, *op. cit.*
76. Jamal, S.H., 'Distribution and types of Beads in the Ganga Valley' *Puratattva* 11, New Delhi, 1979-80, pp. 133-39.
77. Tripathi, 1976, p. 93.
78. *Ibid.*
79. *Ibid.*
80. Dikshit, 1984, *op. cit.,* p. 543.
81. *Ibid.*
82. Bhan, 1975, *op. cit.*
83. *Ibid.*
84. *Ibid.*
85. *Ibid.*
86. *Ibid.*
87. *Ibid.*
88. *Ibid.*
89. *Ibid.*
90. *Ibid.*
91. *Ibid.*
92. *Ibid.*
93. *Ibid.*
94. Phogat, 1972, p. 262.
95. *Ibid.*

CONCLUSION

In this work the past settlement patterns in the district have been analyzed in the light of important variables such as ecology, topography, soil types, land use patterns, hydrology, drainage, water resources, technological know-how and social organization. With the help of these we have tried to assess the extent of change and continuity, and find out the sequence and spread of human habitations from early Harappan to early historical times.

The inception of human habitation could be traced back to the second half of the early Harappan phase coinciding with the last quarter of the third millennium BC and demonstrated by the presence of such remains at nine newly and three previously explored sites in the old alluvium of the district in the process of exploration. Similar remains have also been reported from the other adjoining areas of Rohtak, Jind and Hisar districts in the south and west. This suggests that the early Harappan people in the area of discussion seem to have advanced from the said adjoining areas and thus, these constituted the eastern limit of their area of influence. The small number and size of these sites with simple artifacts point to their being ordinary rural abodes of thatched mud houses harbouring simple and small rural communities subsisting on small-scale farming, extensive pastoral activities and gathering of wild food. The nature of cultural milieu bears similarity with that found in the western areas in this period. This shows that the level of development remained low for the simple reason of their limited potential to negotiate with the then prevailing harsh natural environment. In such unfavourable conditions people realizing their insurmountable constraints forced to resort to some pragmatic methods (demanding nominal application of tool, technology and labour) and suiting the most to the natural conditions to maximise the prospects to survive the adverse environment. And that is what they managed positively as suggested by the occurrence of their settlements in this phase. We have also tried to understand and explain the meaning of the term environment. According to us it is not the natural environment alone rather it includes the natural conditions, technical expertize and social awareness and organization collectively amount to constitute the concept of environment in practical sense of its connotation. Whenever any of the constituents is used in an independent capacity, the relevant constituent variable is used separately, i.e., the technical knowhow, the social organization, etc.

The next stage in the history of human settlement is marked by the occurrence of a somewhat distinct tradition which being associated with the urban Harappan tradition is popularly known as mature Harappan phase. The remains of this tradition have been encountered at 10 new and 7 previously explored sites, which are either superimposed upon the earlier remains or are settled on natural surface. This shows that some of the earlier sites could not retain people in succeeding Harappan times, especially sites in the eastern subzone of the Bhangar. And the Harappan people made their new settlements in the western subzone itself. As a result their area of influence was reduced to only the western subzone of the Bhangar unlike the preceding early Harappan phase when it used to cover both the western and the contiguous eastern subzones of the Bhangar. Nevertheless within the former subzone the Harappan people enlarged their area of influence by breaking new grounds around preceding habitations. This underlines partial overlap in their approach to site selection and relative prevailing instable environment around some of the settlements. However, the obvious upward trend in the number and size of the settlements and quality and quantity of cultural artifacts in the western subzone indicate relatively stable environment around them. They also seem to have used mudbricks as noticed on the surface of Rindhana-I for making their houses. We are not sure about the nature of the operative factors for want of concrete supporting evidence in the absence of excavations in our area. This is, however, sought to be explained with the help of our observations in the field and their verification with the help of the excavated material from the adjoining areas. It is, thus, observed that there was no substantial change in their activity loci and prevailing natural environment from early Harappan to Harappan times. And the reason for evolutionary change lies with the growing social and technological awareness of the people as a general corrollary of their being part and parcel of a general environment of development that was in vogue around in those days. This shows how the accommodative outlook of the early Harappan people experienced a shift towards growing intervention in Harappan times. With the result these people attempted to harness the extant environment according to their needs, albeit on a restricted scale. Nevertheless the nature and character of the people and their settlements in this zone remained primarily rural consisting of small communities depending on mainly agriculture and pastoral activities for their subsistence.

After the Harappan phase the evolutionary process stands reversed in the late Harappan phase for neither of the characteristic features of the preceding Harappan phase was noticed in circulation. Conditions like the early Harappan phase prevailed. But interestingly contrary to the two previous phases the number of settlements multiplied considerably taking their collective numerical strength to sixty and accordingly an increase in their area of influence as they encompassed the whole of the Bhangar area of the district. Even they penetrated the Khadar in some form as explained by the occurrence of such remains from the sole site of Datauli in district Sonipat and Dikadala in Panipat district. This sudden growth of population and the number of settlements on the one hand and the reversal in material culture on the other are seen in the light of the disintegration of the Harappan city system in the west, i.e., in the valleys of the Greater Indus river system. Second interestng feature is the continued predominant position of the western zone of the Bhangar over other parts, like the previous phases, for its settlement strength is still greater than the eastern section of the Bhangar, the Western Margins of the Khadar and the Khadar itself. This is mainly attributed to the balance between the growth of population by addition of incoming people and the carrying capacity of the environment. The moment this balance was affected adversely people visualizing non-tenability of the lifeline began budding off or migrating from the parent sites and the areas of activities. This resulted in the increase of settlements and the area of influence as well. The survival of large number of people in the hitherto inhospital looking zone was perhaps the outcome of their improved operative potential. And this transformation was effected by the new technological know-how brought along by the people coming from the urban centres and by the addition of new subsistence strategies. Third there was a commonality with the previous phases in their approach to site selection, subsistence practices, nature and character of settlements and cultural milieu as their habitations are located on rising pockets, especially aeolian in nature, to subsist by way of small farming and large-scale pastoralism. They also seem to have shifted their settlements more frequently than we could visualize, if we take cognizance of flimsy habitational depositions at a sizeable number of settlements in this phase. It seems that with the disintegration of the city system the impulse to growth and stability was lost and hence, the consequent conditions prevailed.

In the succeeding phase of PGW-GW the number of occupations went up to sixty three and their zonal distribution shows higher number of settlements in the eastern half of the Bhangar than the western zone. For the first time the balance of concentration shifted in favour of the eastern subzone of the Bhangar. The trend of increasing attention to this zone was, of course, recorded for the first time in late Harappan times gets consolidated in the PGW-GW phase to promote it to a higher position. As a result this zone assumes the status of a densely populated zone. Secondly the area along the Western Margins of the Yamuna river, i.e., a dividing line of the Bhangar in the west and the Khadar in the east, received still greater attention than any other zone for there are as many as ten settlements along this line from north to south. This number is very high as compared to the total covered area of this zone. Not only the number of such settlements is higher, but they are closely spaced and enjoy relatively bigger size in terms of area as well as depositional thickness. This was possible because of optimum level of harmony among variables of the environment. This ideal level was provided by a very rich combination of two contrasting geographical units, the Bhangar and the Khadar, which promised comparatively better avenues for subsistence and development than the major zones of the Bhangar and the Khadar themselves in the then available operative potential. This in conjunction with other things being equal paved the way for better growth so obvious in the remains. Third in this very period the remains of settlements were recorded at eight places in the Khadar also. This number of occupations is indicative of the fact of its being occupied in this phase for the first time in a reasonable manner, especially when compared to the preceding late Harappan phase. In terms of time these settlements seem to belong to the last stage of this phase as demonstrated by the inferior cultural equipments especially dull and rough grey ceramic wares usually devoid of any painting designs. This is true to the majority of the explored sites in this section. Fourth the lower half or the southern parts of the Khadar failed to produce any settlement, even in this phase. Keeping in view its low-elevation and the general environment, it is observed, the people failed to make any effort in the face of unpredictable incoming waters of the Yamuna. Fifth the average location of the sites tempts us to equate their site selection process with those of the preceding phases, i.e., usually settlements are located on rising pockets irrespective of their being aeolian or alluvial in origin. Their common approach to site selection is further underlined by the occurrence of PGW-GW remains on (early Harappan, Harappan and late Harappan) many past settlements in the west. Thus, the increase in the number of occupations, population and the area of influence seem to have been the result of the improving environment over the period, especially towards the last stage of this phase people became growingly aware about the minute details of the landscape, the ecological conditions, the behaviour of the river and the methods to harness them to their benefit at least at a low level.

Coming to the early historical period an improved situation in general is attested. However, there is a special increase in the number of habitations, which reached to 88. The shift from the west zone (20) to the east (31) is further consolidated. The Western Margins and the Khadar areas received greater attention of the people to occupy 18 and 19 locations respectively. The spread of these settlements touches almost all the areas. The increase from 8 in PGW-GW levels to 19 in the succeeding RW-NBPW levels is commendable. In this manner a process of inhabitation that began way back in early Harappan times seems to have attained its watermark in the early historical period when the entire area of the district experienced the feel of human activities.

The approach in regard to site selection in early historical period is in conformity with that followed by the people at various stages of development of human settlement history in the past. The locational analysis exhibits a usual craving among the people for elevated pockets coupled with natural depressions. These very depressions being water repositories seem to have formed the main lifeline in the past as well. In the absence of any other sources of water so was the impact on the mind of the people that they settled mainly along the extension of these depressions as explained by the usual orientation pattern of the settlements. However this trend stands reversed in modern times in both the geographical zones. It is caused by the introduction of tap water supply among other considerations in the Bhangar whereas in the Khadar the causative factor is the introduction of new techniques to harness the ground water.

Regarding carrying capacity and spliting up tendency among the inhabitants at the different stages of development we have observed that the carrying capacity of any locale conforms well with the overall prevailing conditions as noticed above. And this is subjected to the level of harmony among the said operative variables. We have also marked, whenever the first half of the environment, i.e., the natural forces, was seen playing a predominant role as was the case during early Harappan, Harappan and late-Harappan levels, the carrying capacity remained on the lower side. And whenever the second half of the variables of the environment, such as level of technological know-how, human organization and awareness, was improved the predominance of the natural forces was correspondingly reduced as was in PGW-GW levels in the Khadar and at some pockets like Bhawar-I, Chhapada, Gangesar-II, Ghadwal-III (in western zone) and Khanpur-I and Rithal (in eastern zone) in the Bhangar. And the improvement in the second half proved still more effective in the PGW-GW phase itself along the Western Margins area for the sites are closely spaced and enjoy better size and cultural deposition. These sites belong to the second stage of development. In the realm of harmonious relations still more advanced stage was also noticed, that is the still higher level of harmonious relations. In this context with the further improvement in the existing levels of awareness, organization and technical expertize with the passage of time the potential of the people to negotiate with the given natural environment enhanced substantially in the succeeding RW-NBPW phase. And this is considered to be the optimum level up to now. Here the word optimum is used in pre-industrial context. Accordingly it denotes a level of development where neither of the constituents of the environment gets affected adversely, which may ultimately influence the very development of society in a negative manner. Hence, the optimum level stands for best possible harmony between and among the operative variables of development. And if we take cognizance of the modern development process in the area the desired harmony is being consistently undermined to pose a serious threat not only to the ecological balance, but also to the very primary source of subsistence, the agriculture, for the optimum level of harmony is threatened by the over exploitation of the natural resources with the help of advanced technological know-how and awareness. This process has affected the growth patterns adversely in agriculture in recent times. This is a point of serious concern these days and the scientists are now resorting to the traditional strategies to arrest any further negative impact on the environment and the prospective growth patterns. Therefore, the imbalance on either side in the precarious level of harmony is liable to affect the pace of production and development accordingly.

In the light of the above observations we marked as long as the harmony is maintained among the catalyst forces the splitting up tendency is reduced to a greater extent. But the moment this is disturbed either way people feel either compelled or tempted to migrate to any other suitable place. We noticed that people in the Bhangar region tended to stay back at the parent settlements till a point beyond which it became unavoidable to part off for the increasingly insufficient supply of food due to the disturbances occurred in the harmonious relations of constituents of the environment. Whereas in the Khadar zone people were forced to resist the temptation till the moment they became sure both of their independent subsistence and existence at a new place. In the RW-NBPW phase people seem to have managed to gather that sense of confidence, which resulted in making of 11 new settlements apart from continuation of habitation on eight sites of the previous period. This trend became more pronounced among the modern villages of this area turning this inhospitable looking zone into a highly populated area with the recent upcoming numerous small, but closely spaced villages. So strong and fast is the tendency to make new settlements in this area that the authorities fail to cope with by putting them on records as we marked that some of the such villages (newly found) are not, even on the record of the district.

We also note the inherent differences in the two natural sources of the Bhangar and the Khadar. The latter despite its being substantially rich in regard to natural resources was avoided for habitational purposes by the people in the early stages of settlement history. Whereas the so-called most unattractive and inhospitable looking Bhangar zone relished by people right from the beginning of this process. We also know in this section water is awfully scarce, which is so abundant in the Khadar. Had water alone been of so paramount importance to make habitations the Khadar area should have been the formative zone for habitations right from the beginning of this process. And the area with precarious water resources should have been avoided completely. And how is that in the subsequent stages (PGW-GW and RW-NBPW) people gradually became inclined to make some settlements and later aspired to occupy the Khadar zone fully? To be specific the simple explanation lies with the qualitative difference between the two zones on the one hand and in the different stages of settlement history on the other. We know that till the late Harappan time the copper technology with its inherent limitations was ineffective in areas like that of the Khadar. This was capable of harmonize well with only dry conditions like those of the Bhangar. Accordingly other variables such as human organization and pragmatic strategies were also suffered from serious limitations. With the result it was neither advisable nor possible to traverse in unknown hostile areas like the Khadar. But in the PGW-GW and RW-NBPW levels these handicaps were substantially eliminated with the growing awareness, organization and technical expertize in the form of iron tools to harmonize with the hitherto inhospitable looking Khadar zone. Thus, it is the combination of operative variables which matters rather than any individual aspect to guide the process of human activities. People usually tend to search and select such conditions where they could negotiate successfully to lead their lives. And that is what we discern among the people during the period.

With regard to the hierarchy among settlements of different periods, we observed a fair degree of variation in their size, depositional elevation, clustering and frequency of artifacts. In the early Harappan and Harappan phases the Rindhana-Ghadwal complex supported higher number of settlements with better cultural equipment. Similarly in the PGW-GW phase good settlements were established at Khanpur Kalan-I, Kakana and Rithal in the eastern half of the Bhangar. Settlements of this type were also located at Gumad, Agwanpur, Sandhal Khurd, Harsana Kalan-I and Akbarpur Barota along the western margins of the Khadar. Settlements like Pipli Khera and Dhaturi in the Khadar zone depict the same situation. In the RW-NBPW levels Rindhana-VI, Ghadwal-IV, Chhapada and Khandrai in the western Bhangar; Rukhi-III, Bali-II, Sargthal, Riwada, Bhadana and Khanpur Kalan-I in the eastern section of the Bhangar; Balandpur Kheri, Agwanpur, Sonipat, Harsana Kalan-II and III, Akbarpur Barota and Chhatehara along the western margins; Panchi-Gujjaran, Datauli, Pipli Khera, Murthal and Joshijat in the Khadar are very important settlements. As these sites are not confined to any particular period and locale their mechanism to gain strength rests with the combine of the natural conditions (as these varies at microlevel) of these site and their respective potential to negotiate with the natural forces. Wherever this adjustment had been better the process of growth got accelerated to produce such variations in settlements. This further underlines the uneven distribution of natural resources and the differential operative potential over the landscape through times. And the combine of the two determined to a greater extent the process of development, generally uneven in accordance with the prevaiting operative potential.

To understand the secret of the continued occupation of the Bhangar belt and the subsequent possession of the Khadar as well we have enquired into their subsistence strategies adopted by the people during the period. As we have already stated that the people belonged to primarily agro-pastoral communities who subsisted on rainfed agriculture as there was no river and favourable hydrological regime. The rainfall pattern was also erratic as it is even in modern times. Their technical expertize being highly limited was not promising enough to make up the existing deficiencies in any appreciable manner, especially in the Harappan times. Even their organization strength was not sufficient in any manner to redress the impared natural conditions beyond a point. In this environment of irrepairable constraints people being handicapped remained constantly exposed to the ensuing vicissitudes of nature. In the face of these challenges people seem to have devised and resorted to some pragmatic approaches of selecting suitable crops, crop systems and soils, which were most suited to the existing environment to raise 'barani' or dry crops of barley, gram and millet. These demanded minimum of labour and water, and minimum preparation of the seed bed which was possible with the help of simple tool techniques. Natural provision, as noticed in the combination of soil of varying elevations and variations in the realm of rainfall-system on microlevel, for fallow system and croprotation ensured food production. The prospects of food production were further enhanced by application of crop-combinations. Such practices were not uncommon among villagers until the introduction of green revolution. And this is how people managed to survive by employing simple, but effective subsistence strategies in harmony with the environment and nature.

Fig. 5.22: Painted Pottery

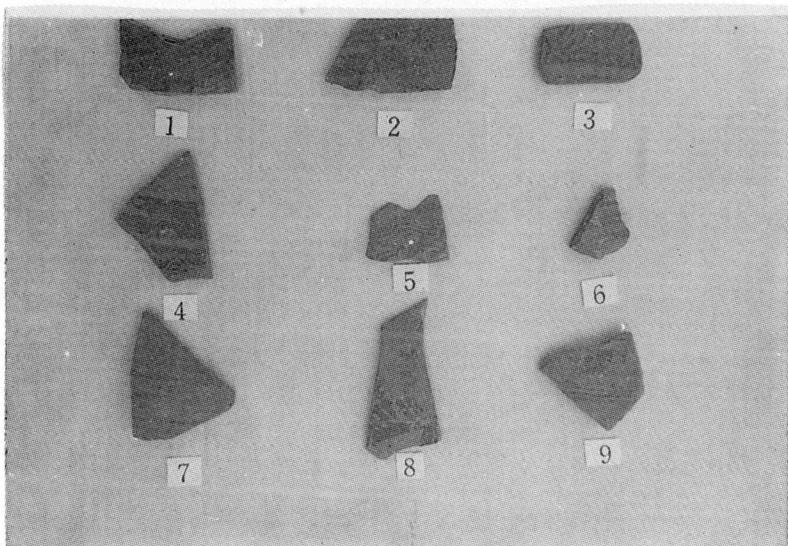

Fig. 5.23: Painted Pottery

Fig. 5.28: Incised Pottery

Fig. 5.29: Incised Pottery

Fig. 5.30: Incised Pottery

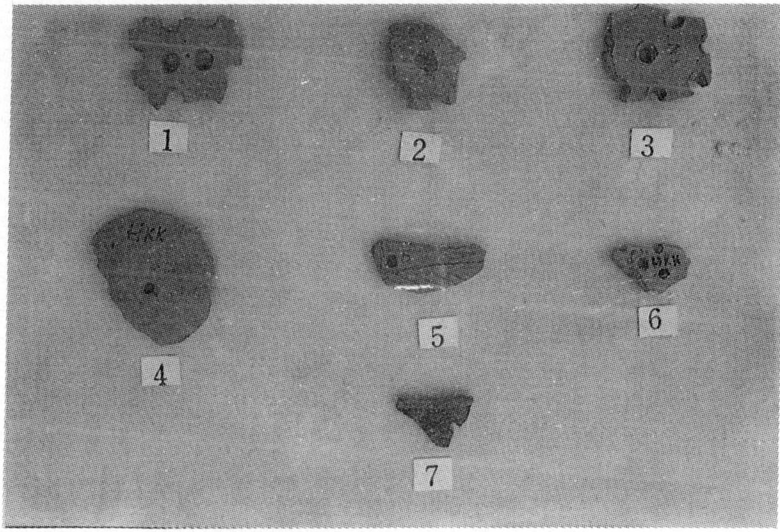

Fig. 5.31: Perforated Pottery

Fig. 5.32: Incised Pottery

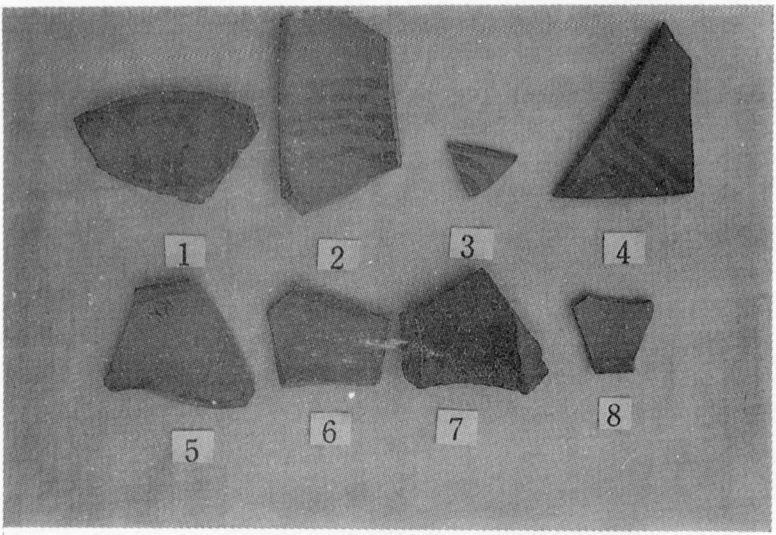

Fig. 5.33: Painted Grey Ware

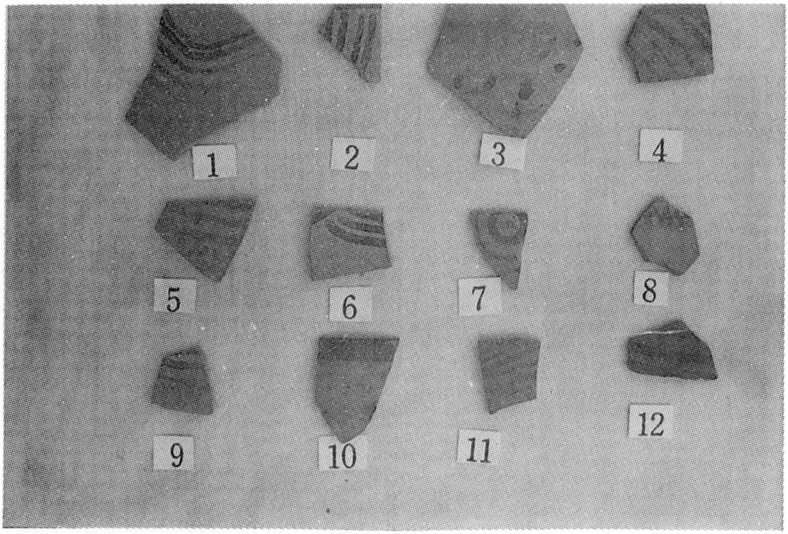

Fig. 5.34: Painted Grey Ware

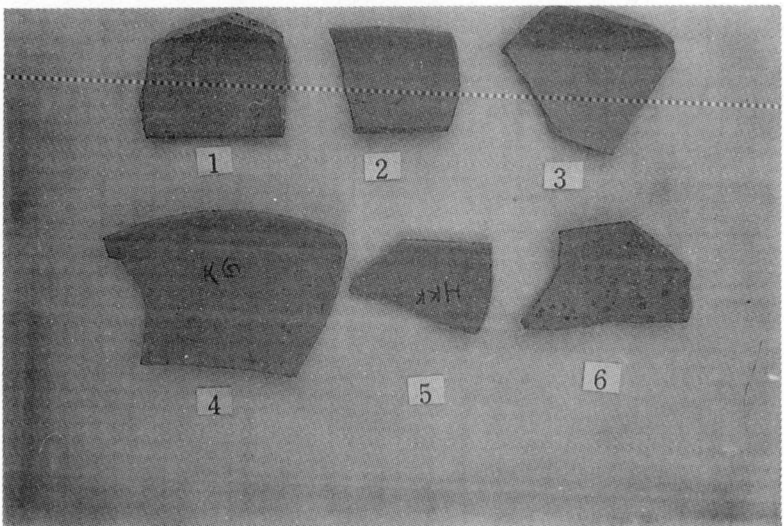

Fig. 5.35: Painted Grey Ware

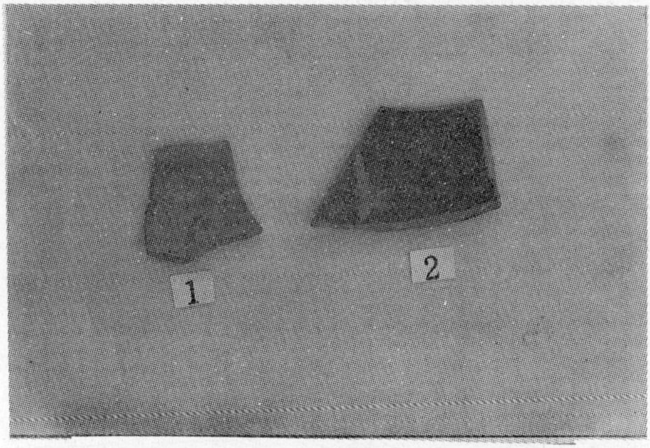

Fig. 5.36: Painted Grey Ware

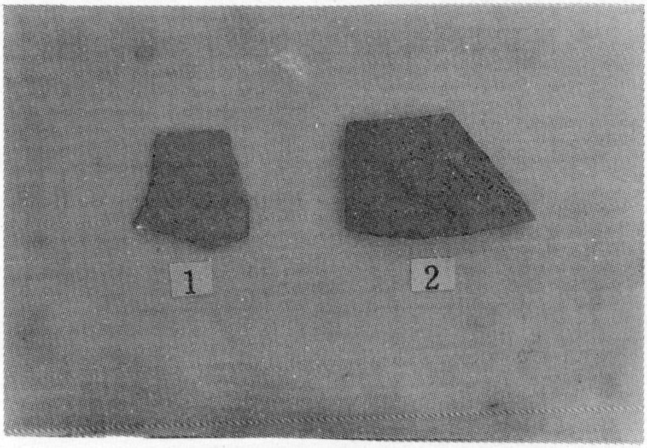

Fig. 5.37: Painted Grey Ware

Fig. 5.38: Incised Grey & Black Slipped Ware

Fig. 5.39: Grey & Black Slipped Ware

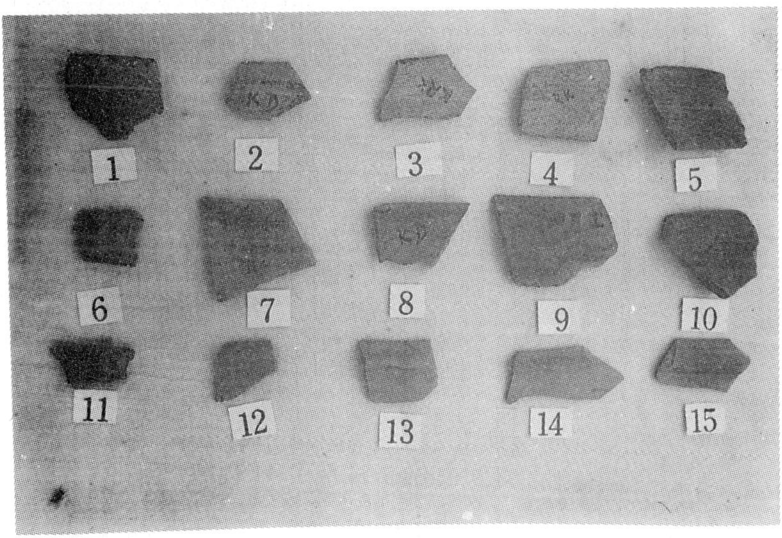

Fig. 5.40: Grey Ware

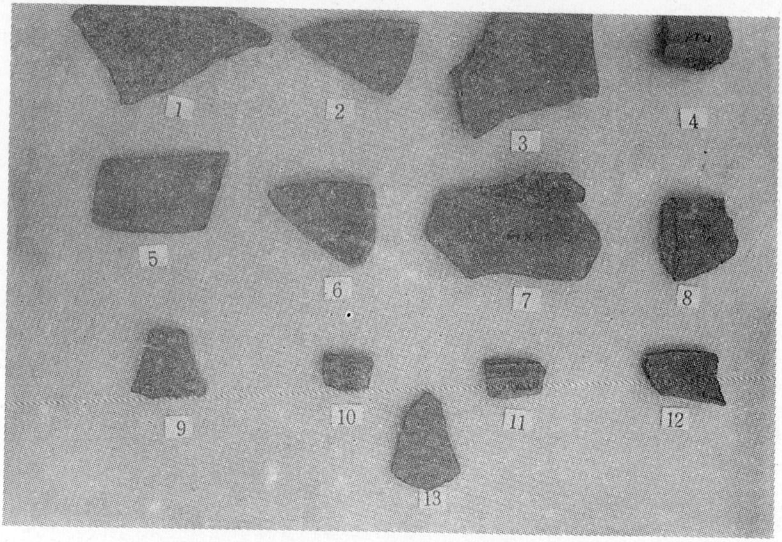

Fig. 5.41: Grey Ware

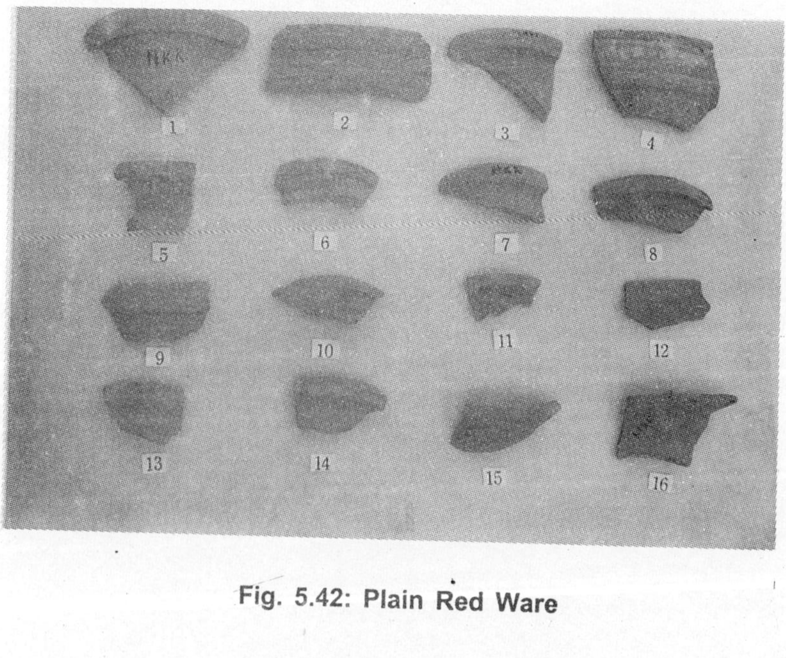

Fig. 5.42: Plain Red Ware

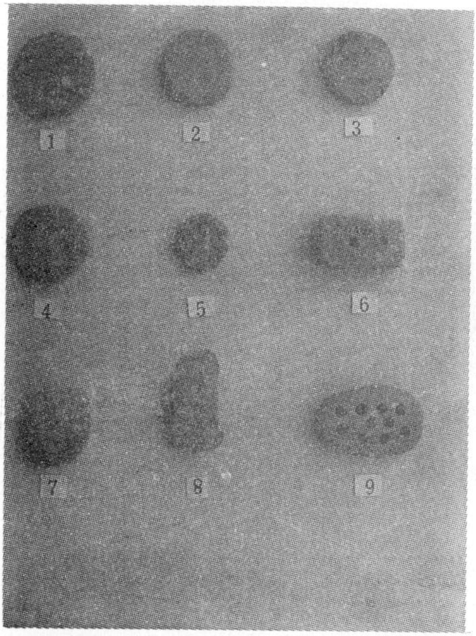

Fig. 5.43: Terracotta Discs

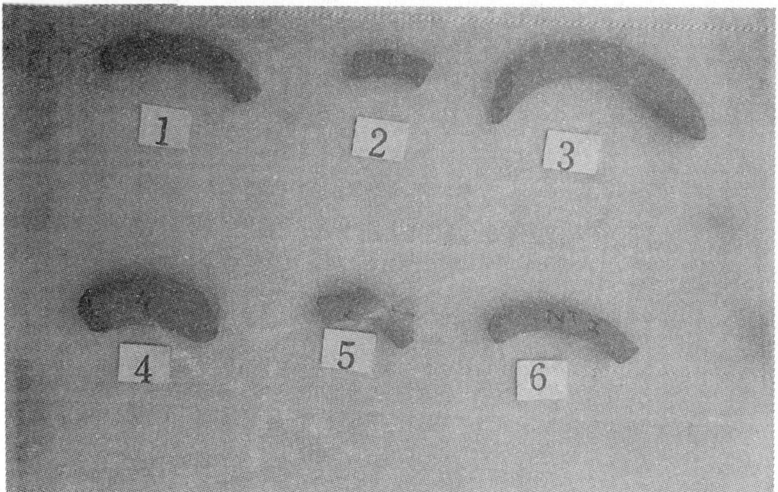

Fig. 5.44: Terracotta Bangles

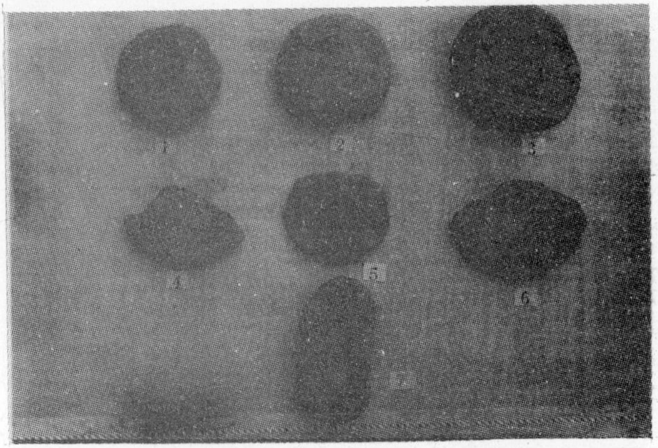

Fig. 5.45: Terracotta Cakes

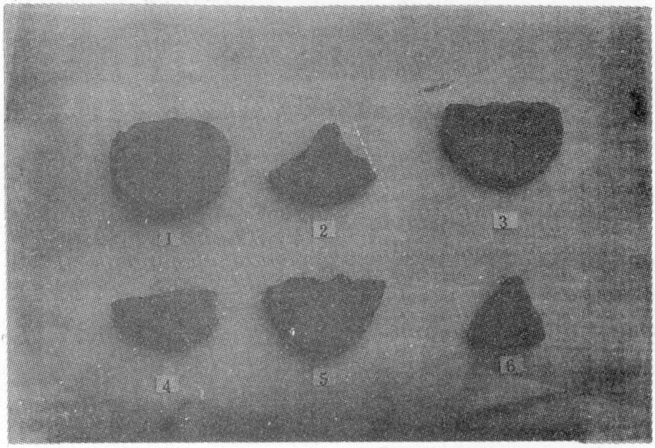

Fig. 5.46: Terracotta Cakes

Fig. 5.47: Terracotta Cakes

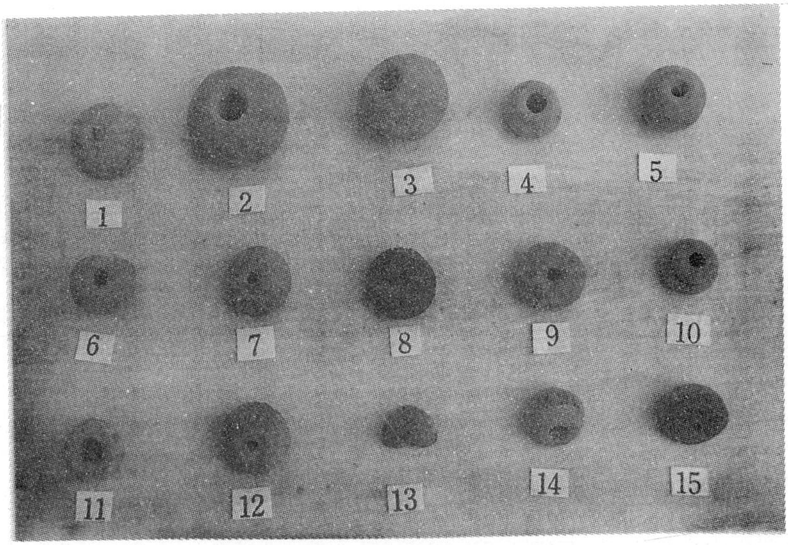

Fig. 5.48: Terracotta Beads

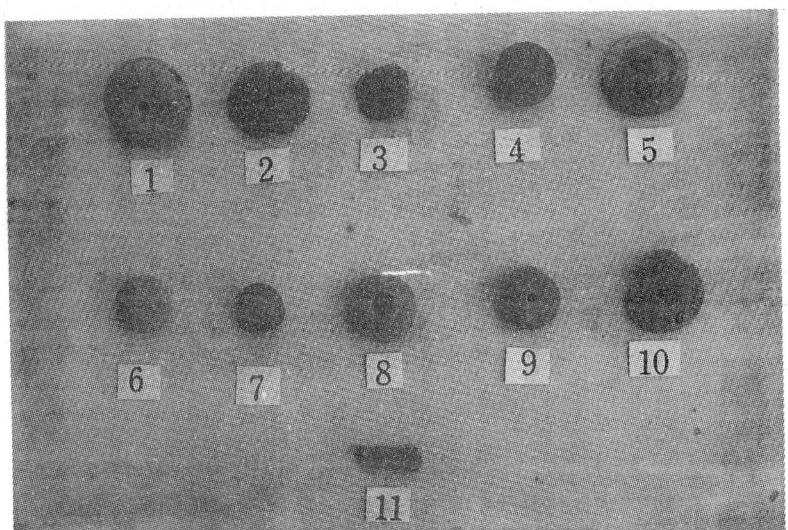

Fig. 5.49: Terracotta Beads

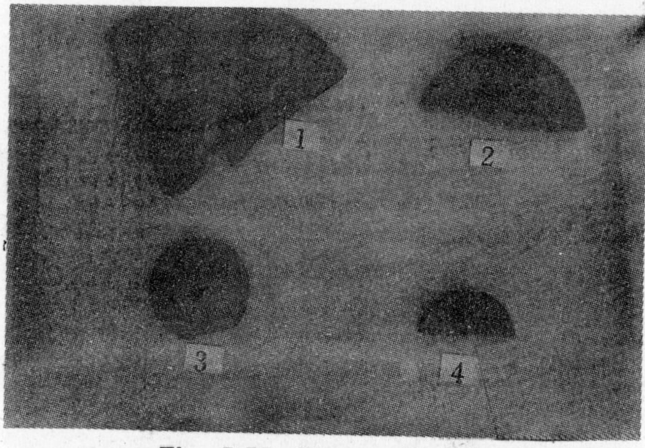

Fig. 5.50: Terracotta Wheels

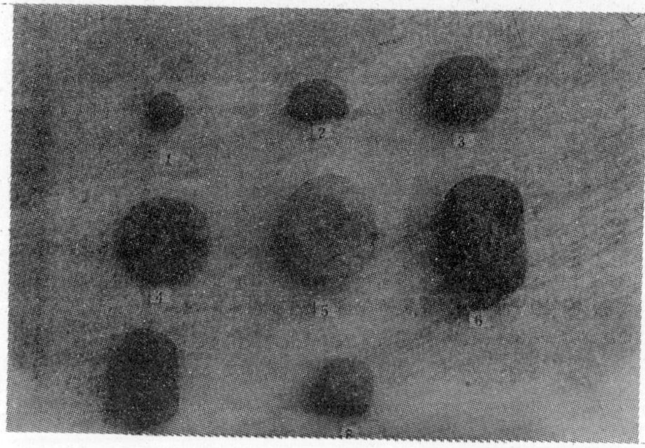

Fig. 5.51: Stone Objects

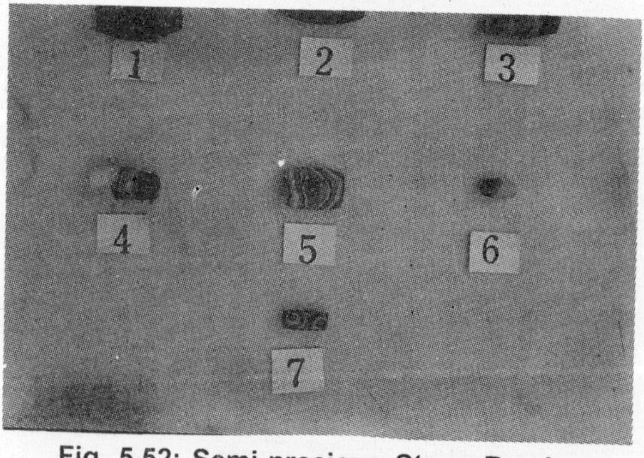

Fig. 5.52: Semi-precious Stone Beads

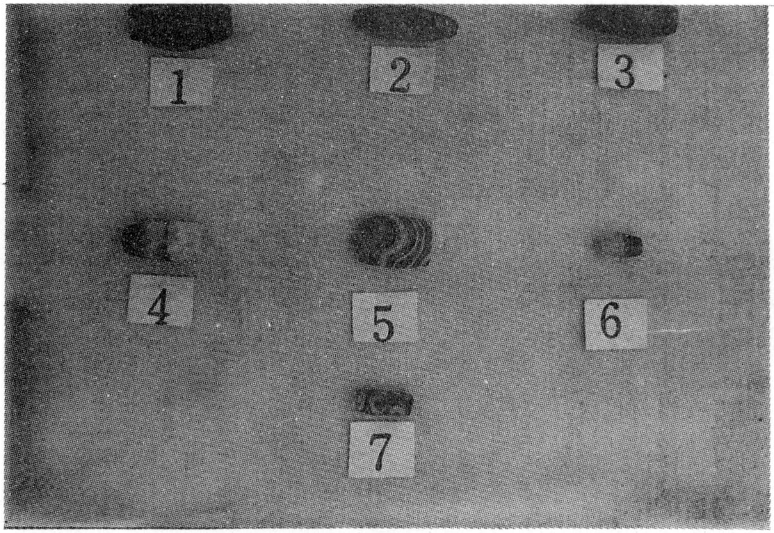

Fig. 5.53: Semi-precious Stone Beads

Fig. 5.54: Village Pond & Well

Fig. 5.55: Village Pond & Well

Fig. 5.56: Village Pond & Well

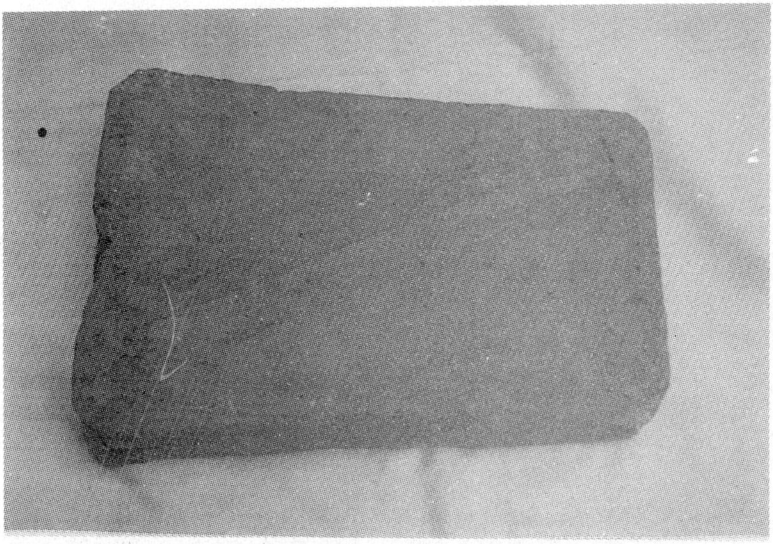

Fig. 5.57: Burnt Bricks

Fig. 5.58: Burnt Bricks

Fig. 5.59: Burnt Bricks

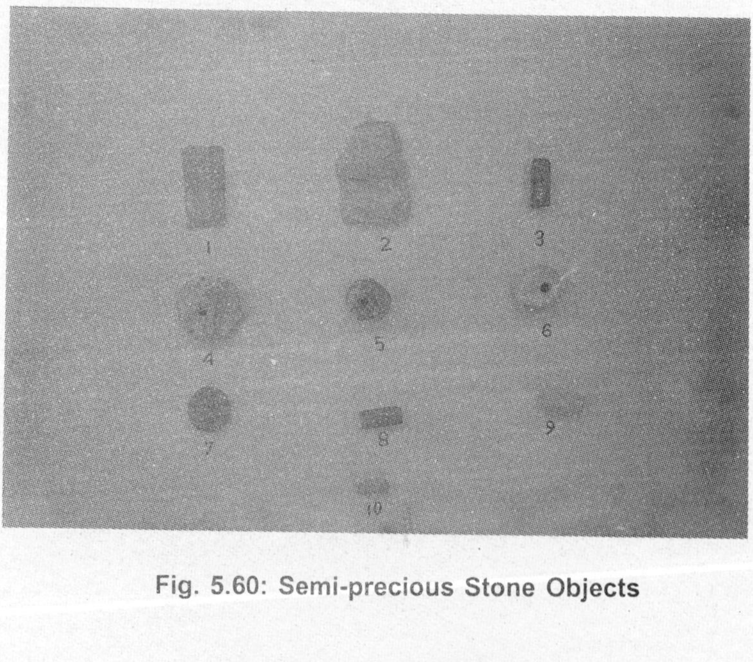

Fig. 5.60: Semi-precious Stone Objects

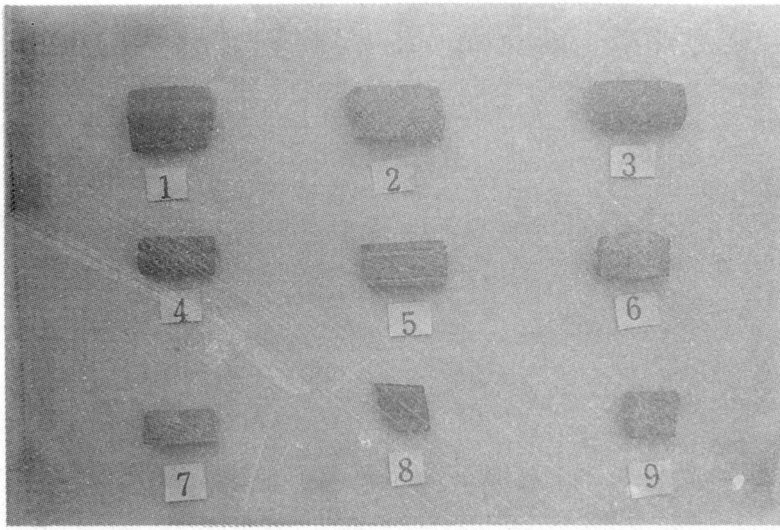

Fig. 5.61: Faience Objects

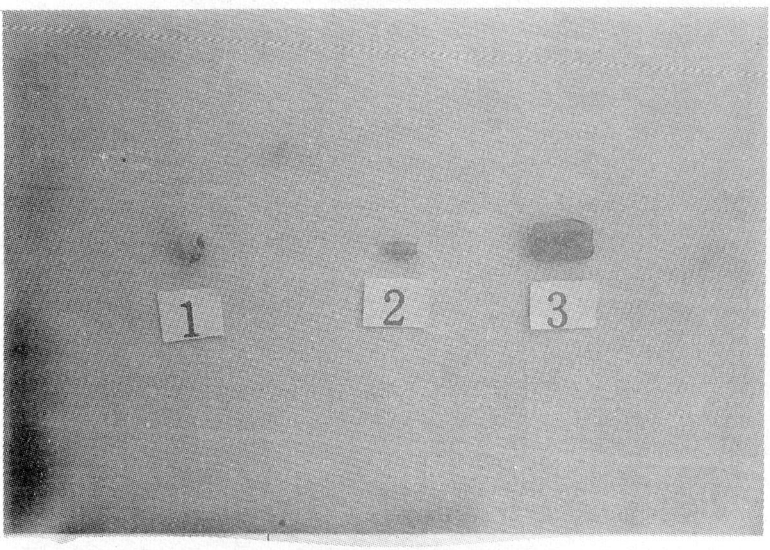

Fig. 5.62: Faience Objects

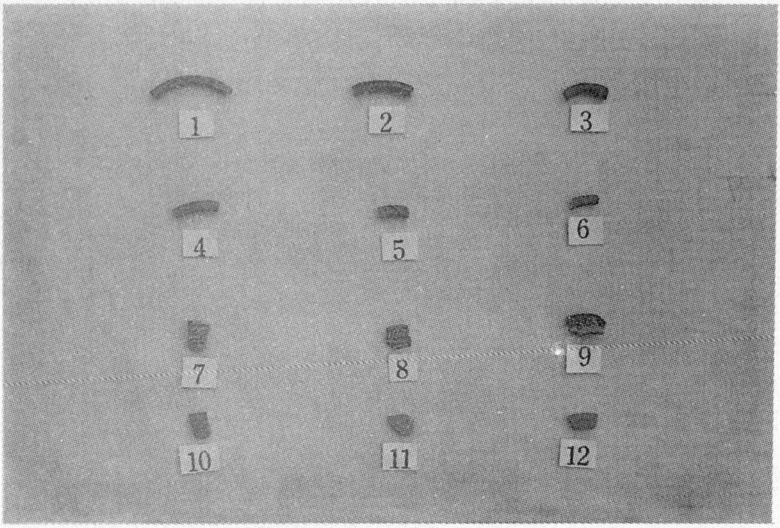

Fig. 5.63: Faience Objects

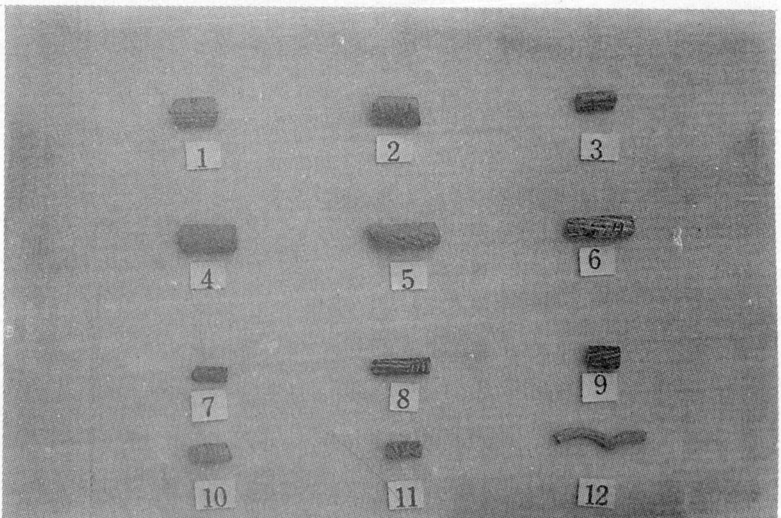

Fig. 5.64: Faience Objects

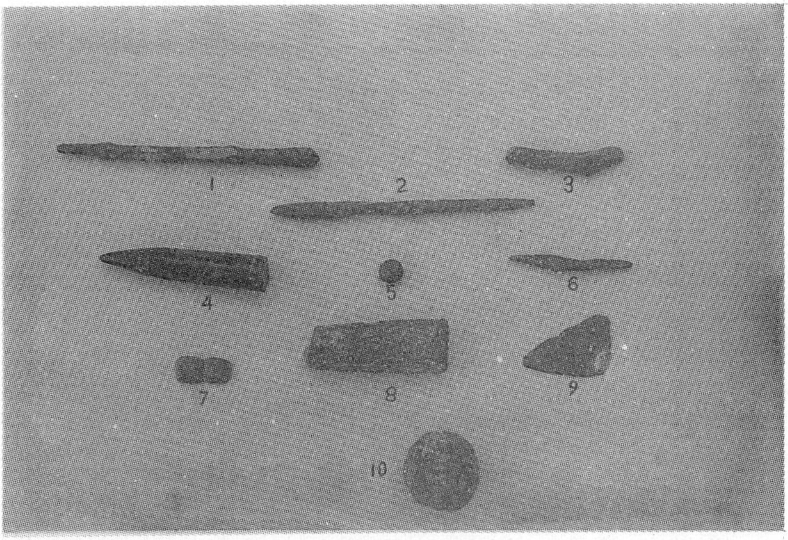

Fig. 5.65: Copper Objects

Fig. 5.66: Miscellaneous Objects

Fig. 5.71: Harappan Painted Pottery

Fig. 5.72: Harappan Pottery & Terracotta Objects

APPENDIX–I

Ancient Sites in the Western zone of the Bhangar-Western Gohana Tahsil

1.	Ahulana-I	EH+H+LH
2.	Ahulana-II	LH
3.	Ahulana-III	LH+PGW-GW
4.	Banwasa	LH
5.	Baroda-I	H+LH
6.	Baroda-II	EH+H+LH+PGW-GW
7.	Baroda-III	H+LH+PGW-GW+RW*
8.	Bhadautikhas	RW+Med.
9.	Bhainswan Khurd	LH+PGW-GW+RW
10.	Bhawad-I	LH+PGW-GW+RW
11.	Bhawad-II	H+LH
12.	Bichpadi-I	LH
13.	Bichpadi-II	PGW-GW
14.	Busana Kalan	LH-OCP+GW
15.	Butana-I	LH+GW
16.	Butana-II	H+LH
17.	Chhapada	H+LH+PGW-GW+BSW+RW-NBPW
18. (17a)	Dhanana	PGW-GW
19.	Gangana-I	LH
20.	Gangana-II	H+LH
21.	Gangesar-I	EH+H+LH
22.	Gangesar-II	LH+PGW-GW+BSW
23.	Ghadwal-I	EH+H+LH

* First letter of each culture abbreviation represents respective cultures in map 4.

24.	Ghadwal-II	H+LH
25.	Ghadwal-III	EH+H+LH+PW-GW+BSW
26.	Ghadwal-IV	RW+Kushan+E. Med.
27.	Ghilaud Kalan	LH+GW+RW+E. Med.
28.	Ghilaud Khurd	LH+PGW-GW
29.	Gohana-I	LH-OCP
30.	Gohana-II	PGW-GW
31.	Isapur Kheri	LH+GW
32.	Jagasi	RW+E. Med.
33.	Jawahara-I	LH+PGW-GW
34.	Jawahara-II	RW
35.	Kahalapa-I	LH
36.	Kahalapa-II	H+LH-OCP
37.	Kathura-I	H+LH
38.	Kathura-II	RW
39.	Khandrai	RW+E.Med.
40.	Mahmoodpur	RW+E.Med.
41.	Mundlana	RW
42.	Nizampur	GW+RW
43.	Nuran Khera-I	H+LH
44.	Nuran Khera-II	LH+PGW-GW+BSW
45.	Nuran Khera-III	EH+LH
46.	Rindhana-I	EH+H+LH
47.	Rindhana-II	EH+H+LH
48.	Rindhana-III	EH+LH
49.	Rindhana-IV	LH
50.	Rindhana-V	PGW-GW
51.	Rindhana-VI	RW+E.Med.
52.	Rukhi-I	EH
53.	Rukhi-III	RW+E.Med.
54.	Siwana Mal-I	PGW+GW+RW
55.	Siwana Mal-II	GW+RW
56.	Thuska	PGW-GW+BSW+RW-NBPW

Ancient Sites in The Eastern Bhangar

1.	Anwali	RW+E.Med.
2.	Badwasani	RW+E.Med.

3.	Baghru	EH
4.	Bajana Kalan	RW+E. Med.
5.	Bali-I	LH+PGW
6.	Bali-II	RW+Kushan+Med.
7.	Barota	LH+PGW-GW+RW
7a.	Bhadana	RW+E. Med.
8.	Bhambad	RW+E. Med.
9.	Bhatgaon-I	RW+E. Med.
10.	Bhatgaon-II	LH+GW+RW
11.	Bhatgaon-III	LH
12.	Bidhal	LH-OCP
13.	Bilbilan	RW+E. Med.
14.	Bohala	LH+PGW+GW
15.	Chatia Deva	LH+PGW-GW
16.	Dodwa	LH+PGW+GW
17.	Farmana-I	PGW+BSW
18.	Farmana-II	EH+LH+PGW+GW
18a.	Garhi Bala	RW+E. Med.
19.	Gorad	PGW-GW
20.	Halalpur-I	PGW-GW
21.	Halalpur-II	EH+LH+RW+E. Med.
22.	Jasrana	RW+E. Med.
23.	Kailana Khas	RW+E. Med.
24.	Kakana	LH+PGW-GW+RW
25.	Kakroi	RW+E. Med.
26.	Khanpur-I	LH+PGW-GW+RW
27.	Khanpur-II	LH
28.	Kharkhauda	RW+E. Med.
29.	Kundal-I	RW+E. Med.
30.	Kundal-II	RW+E. Med.
31.	Mandori	RW+E. Med.
32.	Mohana	RW+E. Med.
33.	Nakloi-I	PGW+BSW
34.	Nakloi-II	GW+RW
35.	Pipli-I	LH-OCP
36.	Pipli-II	RW+E. Med.
37.	Rithal	LH+PGW+GW+BSW

38.	Riwada	RW+E. Med.
39.	Rolad Kalan	RW+E. Med.
40.	Rukhi-II	LH+OCP+GW
41.	Sargthai	LH+OCP+GW
42.	Samdi	LH+OCP+GW+RW
43.	Silana	EH
44.	Sisana-I	LH+GW
45.	Sisana-II	RW+E. Med.
46.	Thana Khurd	RW
47.	Zhinjhauli	RW

Ancient Sites Along The Western Margins

1.	Agwanpur	GW+RW+Med.
2.	Akbarpur Barota	PGW+GW+BSW+RW-NBPW
3.	Balandpur Kheri	PGW-GW+RW+Med.
4.	Chatia Aulia	RW+RW
5.	Chhatehara	RW+E. Med.
6.	Gujjar Kheri	RW
7.	Gumar	PGW+BSW+RW-NBPW
8.	Harsana Kalan-I	LH+PGW-GW
9.	Harsana Kalan-II	RW+E.Med.
10.	Harsana Kalan-III	RW
11.	Harsana Kalan-IV	RW+E Med.
12.	Jawahari-I	PGW-GW
13.	Jawahari-II	RW+E.Med.
14.	Laharada-I	LH
15.	Laharada-II	GW+RW+E.Med.
16.	Laharada-III	RW+Med.
17.	Nahari-I	RW+E.Med.
18.	Nahari-II	LH+RW+E.Med.
19.	Sandhal Khurd	GW+RW
20.	Shahzadpur	RW+E.Med.
21.	Sonipat	PGW-GW+RW-NBPW

Ancient Sites in the Khadar

1.	Bhagan	GW+E.Med.
2.	Chirsami	GW+GW+E.Med.

3.	Datauli	LH+PGW-GW+GW+E.Med.
4.	Dhaturi	PGW+GW+RW+Gupta
5.	Ganaur	RW+Med.
6.	Ibrahimpur Kurad	RW
7.	Jakhauli	RW+E.Med.
8.	Joshi Jat	RW+E.Med.
9.	Kami	RW+E.Med.
10.	Kheri Tega	RW+E.Med.
11.	Murthal	RW+E.Med.
12.	Nangal Khurd	GW+RW
13.	Panchi Gujjran	GW+RW
14.	Pipli Khera	GW+BSW+RW
15.	Qamaspur	GW+BSW+RW+NBPW
16.	Rewali	GW+RW-NBPW
17.	Sandhal Kalan	RW+E.Med.
18.	Sewali	RW+E.Med.
19.	Tajpur	RW+E.Med.

APPENDIX–II

List of Single or Major Culture Sites

Sl. No.	Site Name	Tehsil	Culture*	Area (sq.m.)**	Estimated Population
Early Harappan					
1.	Ahulana-I	Gohana	EH+H+LH	990	20
2.	Gangesar-I	"	EH+H+LH	2020	40
3.	Ghadwal-I	"	EH+H+LH	2060	40
4.	Ghadwal-III	"	EH+H+LH	990	20
5.	Rindhana-I	"	EH+H+H+LH	2020	40
6.	Rindhana-III	"	EH+LH	2325	45
7.	Rukhi-I	"	EH	2040	40
Harappan Sites					
8.	Baroda-I'	"	H+LH	3000	60
9.	Baroda-III	"	H+LH+PGW+RW	4020	80
10.	Bhawad-II	"	H+LH	12000	240
11.	Butana-II	"	H+LH	1090	200
12.	Gangana-II	"	H+LH	4020	80
13.	Ghadwal-I	"	EH+H+LH	20060	400
14.	Ghadwal-II	"	H+LH	16000	300
15.	Kathura-I	"	H+LH	4020	80
16.	Rindhana-II	"	EH+H+LH	990	20
Late Harappan Sites					
17.	Ahulana-II	Gohana	LH	990	20
18.	Banwasa	"	"	990	20
19.	Baroda-II	"	PH+H+LH+PGW	4020	80
20.	Bhainswan Khurd	"	LH+PGW+RW	4020	80

21.	Butana-I	Gohana	LH+GW	990	20
22.	Chhapara	"	H+LH+PGW +BSW+RW-NBP	4020	80
23.	Gangana-I	"	LH	4020	80
24.	Ghilaur Khurd	"	LH+PGW	4335	90
25.	Isapur Kheri	"	LH+PGW	4020	80
26.	Jawahara	"	LH+PGW	4020	80
27.	Kahalapa-I	"	LH	4918	100
28.	Kahalapa-II	"	H+LH	4020	80
29.	Khathura-I	"	H+LH	4020	80
30.	Nurankhera-I	"	H+LH	4020	80
31.	Nurankhera-II	"	LH+PGW	4020	80
32.	Nurankhera-III	"	PH+LH	4020	80
33.	Rindhana-IV	"	LH	8146	165
34.	Bali	"	LH+PGW	2040	80
35.	Barota	"	LH+PGW+RW	4335	90
36.	Bidhal	Sonipat	LH	990	20
37.	Bohala	"	LH+PGW	4020	80
38.	Chatia Deva	"	LH+PGW	5257	105
39.	Khanpura-I	Gohana	LH+PGW+RW	8134	165
40.	Khanpura-II	"	LH	990	20
41.	Pipli-I	Sonipat	LH	2040	80
42.	Rithal	Gohana	LH+PGW	4337	90
43.	Rukhi-II	"	LH+PGW	2040	80
44.	Sisana-I	Sonipat	LH+GW	1990	40
45.	Laharada-I	"	LH	9236	185
46.	Nahari-II	"	LH	1990	40

PGW Sites

47.	Bhawad-I	Gohana	LH+PGW	16187	325
48.	Bichpari-II	"	PGW	2432	50
49.	Gangesar-II	"	LH+PGW	.14020	280
50.	Rindhana-V	"	PGW	4020	40
51.	Siwana Mal-I	"	PGW+RW	4020	40
52.	Farmana-I	Sonipat	PGW+BSW	4020	40
53	Farmana-II	"	LH+PGW	1800	160
54.	Gorad	"	PGW	4020	40

55.	Halapur-I	Sonipat	PGW	4020	40
56.	Kakana	Gohana	LH+PGW	18134	365
57.	Khanpur Kalan	"	LH+PGW+RW	18134	365
58.	Sisana-II	Sonipat	LH+PGW	4020	80
59.	Akbarpur Barota	"	PGW+BSW+NBP	18134	365
60.	Gumar	"	PGW+BSW+NBP	24258	485
61.	Harsana Kalan-I	"	LH+PGW	18133	365
62.	Jawahari	"	PGW	8073	160
63.	Sandhal Khurd	"	GW+RW	18136	365
64.	Dhaturi	"	PGW	8134	165
65.	Piplikhera	"	PGW+BSW+RW	12123	245
66.	Qamaspur	"	GW	12137	245

RW-NBPW Sites

67.	Bhadauti Khas	Gohana	RW+Med.	17030	340
68.	Ghadwal-IV	"	RW+Med.	40962	820
69.	Jagasi	"	RW+E. Med.	40962	820
70.	Jawahara-II	"	RW	18980	380
71.	Khandarai	"	RW+E. Med.	32218	645
72.	Mundlana	"	RW+E. Med.	48668	970
73.	Mahmudpur-I	"	RW+E. Med.	38127	720
74.	Nizampur	"	GW, RW	22138	445
75.	Rindhana-VI	"	RW+E. Med.	36142	720
76.	Rukhi-III	"	RW+E. Med.	24227	485
77.	Thuska	"	PGW+BSW+RW	24563	490
78.	Bhadana	Sonipat	RW+E. Med.	36138	720
79.	Bilbilan	"	RW+Med.	27336	545
80.	Garhi Bala	"	RW+Med.	16236	325
81.	Jasarana	"	RW	32336	645
82.	Kundal-I	"	RW	32136	645
83.	Mandori	"	RW	22421	450
84.	Mohana	"	RW+E. Med.	24563	450
85.	Moi	"	RW	42134	842
86.	Purkhas	"	RW	24225	465
87.	Thana Khurd	"	RW+Med.	27186	542
88.	Agwanpur	"	GW+RW+Med.	27736	516
89.	Harsana Kalan-II	"	RW+E. Med.	27236	545

90.	Harsana Kalan-III	Sonipat	RW	32122	642
91.	Kakroi	"	RW+E. Med.	17500	350
92.	Laharada-III	"	RW+E. Med.	38136	722
93.	Joshi Jat	"	RW+E. Med.	38236	725
94.	Murthal	"	RW+E. Med.	63236	1215
95.	Sewali	"	RW+E. Med.	54136	1082
96.	Tajpur	"	RW+E. Med.	36327	727

* EH—Early Harappan, H—Harappan, LH—Late Harappan, PGW—Painted Grey Ware, GW—Grey Ware, BSW—Black Slipped Ware, RW—Red Ware of early historical Period, NBPW—Northern Black Polished Ware, E. Med.—Early Medieval, Med.—Medieval.

** Wherever double or multiculture remains are witnessed at a site the calculation of their respective area of influence is based on the relative frequency of their artifacts in a portion of the sites. Hence, the figures of area are based on tentative estimates and may differ from the actual recorded area of a site.

BIBLIOGRAPHY

Primary Sources

Corpus Inscriptions Indicarum, Vol. I, 1963.

Geographical Survey of India Publication, 1976.

Interim Report on the Excavation Carried out at Dhermajra, Calcutta, 1952.

Memoirs of Archaeological Survey of India, No. 51, New Delhi, 1936.

Memoirs of Department of Agriculture in India, Botanical Series-8, Calcutta, 1916.

Secondary Sources

Adams, R.M.: *Land Behind Baghdad: History of Settlements on the Diyala Plains*, Chicago, 1965.

————: *Heartland of Cities*, Chicago, 1981.

Agrawal, D.P.: *The Copper Bronze Age in India*, Delhi, 1971.

————: *The Archaeology of India*, London, 1982.

Agrawal, D.P. and A. Ghosh (ed.): *Radiocarbon and Indian Archaeology*, Bombay, 1973.

Agrawal, D.P. and B.M. Pande (ed.): *Ecology and Archaeology*, 1976.

Agrawal, D.P. and D.K. Chakrabarti (ed): *Essays in Indian Protohistory*, Delhi, 1979.

Agrawal, R.R. et al.: *Saline and Alkali Soils of India*, ICAR, New Delhi, 1982.

Agrawal, V.: *Panini Kalin Bharatavarsh* (Hindi), Varanasi, 1969.

Allchin, B. (ed.): *South Asian Archaeology*, Cambridge, 1984.

Banerjee, N.R.: *Iron Age in India*, Delhi, 1965.

Bharadwaj, H.C.: *Aspects of Ancient Indian Technology*, Delhi, 1979.

Bharadwaj, K.: *Irrigation in India: An Alternative Perspective*, ICSSR, New Delhi, 1990.

Bhattacharya, D.K.: *Ecology and Social Formations in Ancient History*, New Delhi, 1990.

Binford, S.R. and L.R.: *New Perspectives in Archaeology*, Chicago, 1968.

Chakrabarti, D.K.: *A History of Indian Archaeology*, New Delhi, 1988.

————: *The Early Use of Iron in India*, Delhi, 1992.

Chang, K.C. (ed.): *Settlement Patterns in Archaeology*, Polo Alto, National Press, 1968.

Chatterjee, S.P. and T.B. Lahiri: *Man and Environment—A Trend Report Survey of Research in Geography*, ICSSR, New Delhi, 1972.

Clark, D.L. (ed.): *Spatial Archaeology*, London, 1972.

Clark, G.: *World Prehistory in New Perspective*, London, 1977.

Clark, G. and S. Piggott: *Prehistoric Societies*, Harmondsworth, 1976.

Dani, A.H. (ed.): *Indus Civilization: New Perspectives*, Islamabad, 1982.

Deo, S.B. (ed.): *Archaeological Congress and Seminar Papers*, Nagpur, 1972.

Deo, S.B. and M.K. Dhavalikar (ed.): *Studies in Indian Archaeology*, Bombay, 1985.

Dikshit, K.N. (ed.): *Archaeological Perspective in India, Since Independence*, Delhi, 1985.

Duggal, S.L.: *Soil Geographical Zones of the Haryana*, Hissar Agriculture University, 1970.

————: *Soil Geography of Haryana*, Hissar Agriculture University, 1975.

Erdosy, G.: *Urbanization in Early Historic India*, Oxford, 1988.

Flannery, K.V. (ed.): *The Early Mesopotamian Village*, London, 1976.

Gaur, R.C.: *Excavations at Atranjikhera*, Delhi, 1983.

Ghosh, A.: *The City in Early Historical India*, Simla, 1973.

Gupta, S.P. and K.S. Ramchandran.: *Mahabharata, Myth and Reality, Differing Views*, Delhi, 1976.

Hammond, Norman (ed.): *South Asian Archaeology*, London, 1973.

Hartel, Harbert (ed.): *South Asian Archaeology*, Berlin, 1979.

Hodges, H.: *Technology In The Ancient World*, New York 1974.

Hutchinson, S.J. (ed.): *Evolutionary Studies in World Series: Discovery and Change in Indian Subcontinent*, Cambridge, 1974.

Kosambi, D.D.: *The Culture and Civilization of Ancient India in Historical Outline*, New Delhi (Reprint), 1991.

————: *An Introduction to the Study of Ancient Indian History*, (Reprint), Bombay, 1990.

Lal, B.B. *India 1947-1997: New Light on The Indus Civilization*, New Delhi, 1998.

Lal, B.B. and S.P. Gupta (ed.): *Frontiers of Indus Civilization*, Delhi, 1984.

Lal, M.: *Rise of Civilization in Ganga-Yamuna Doab*, Delhi, 1984.

Lamberg-Karlovsky, C.C.: *The Ancient Civilizations in the Near East and Mesopotamia*, London, 1979.

Law, B.C.: *Ancient Geography of India*, Delhi, 1976.

Lec, B.R. and Irveni De Aldine (ed.): *Man The Hunter*, Vone, 1968.

Limbrey, Susan: *Soil Science and Archaeology*, London, 1975.

Malik, S.C.: *Indian Civilization: The Formative Period*, Simla, 1968.

Marshall, Sir Johan: *Mohanjodaro and the Indus Civilization*, London, 1931.

Pande, B.M. and B.D. Chattopadhyaya (ed.): *Archaeology and History*, Delhi, 1987.

Phogat, S.R.: *Inscriptions of Haryana*, Kurukshetra, 1978.

Possehl, G.L. (ed.): *Ancient Cities of Indus*, New Delhi, 1979.

Possehl, G.L. (ed.): *Indus Civilization in Saurashtra*, Delhi, 1980.

———: *Harappa Civilization—A Contemporary Perspective*, Delhi, 1982.

Randhava, M.S.: *A History of Agriculture in India*, Vol. I, ICAR, New Delhi, 1981.

Raychaudhury, S.P. et.al.: *Soils of India*, ICAR, New Delhi, 1963.

Reed, C.A. (ed.): *The Origins of Agriculture*, Morton, 1978.

Renfrew, Colin: *The Explanation of Culture Change—Models in Prehistory*, Duckworth, 1973.

Roelofs, R.T. et.al. (ed.): *Environment and Society*, New Jersey, 1974.

Roy, T.N.: *The Gangetic Civilization*, New Delhi, 1983.

Schiffer, M.B. (ed.): *Advances in Method and Theory in Archaeology*, London, 1919.

Shaffer, Jim G.: *Prehistoric Balluchistan*, Delhi, 1959.

Shanin, T. (ed.): *Peasant and Peasant Society*, Harmondsworth, 1971.

Sharma, R.S.: *Material Culture and Social Formations in Ancient India*, Delhi, 1983.

Shrimati, K.M. (ed.): *Indian Archaeology* Since Independence, Delhi 1996.

———: *Reason and Archaeology*, 1998.

———: *Urban Decay in India*, New Delhi, 1987.

Singh, J.: *An Agricultural Geography of India*, Kurukshetra, 1976.

Singh, U.V. (ed.): *Archaeological Congress and Seminar Papers 1972*, Kurukshetra, 1976.

Sinha, B.P.: *Potteries in Ancient India*, Patna, 1969.

Spate, O.H.K.: *India and Pakistan*: London, 1967.

Spodek, Howard and D.M. Sriniwasan (ed.): *Urban form and Meaning in South Asia: The Shaping of Cities from Prehistoric to Precolonial Times*, London, 1993.

Steward, J.H.: *Theory of Culture Change*, Urbana, 1955.

Suraj Bhan: *Mitathal Excavations (1968) and Other Explorations in the Sutlej-Yamuna Divide, Kurukshetra University*, Kurukshetra, 1975.

Subbarao, B.: *The Personality of India*, Baroda, 1958.

Swedland, C.A. (ed.): *Population Studies in Archaeology and Biological Anthropology: A Symposium for American Archaeological Memoir,* 1982.

Taddei, M. (ed.): *South Asian Archaeology*, Napels, 1979.

Thakur, V.K.: *Urbanization in Ancient India*, Delhi, 1981.

———: *Social Dimensions of Technology: Iron in Early India c. 1300-200 B.C.*, Delhi, 1993.

Tonnies, F.: *Community and Society*, Michigan, 1957.

Trigger, B.: *Beyond History: The Methods of Prehistory*, New York, 1968.

Triggar, B.G.: *A History of Archaeological Thought*, Cambridge, 1989.

Tripathi, V.: *The Painted Grey Ware-An Iron Age Culture of Northern India*, Delhi, 1976.

Uccko, P.J. et. al.: *Man, Settlement and Urbanism*, Duckworth, 1972.

Wheeler, Sir Mortiner: *The Indus Civilization*, Cambridge, 1968.

Willey, G.R.: *Prehistoric Settlements in Viru Valley*, 1953.

———: *Prehistoric Settlement Patterns in New World*, Viking Fund Publication in Anthropology, No. 23, 1956.

Wolley, Sir Leonard: *The Beginnings of Civilization: History of Mankind, Cultural and Scientific Development*, Vol. I, New York, 1963.

Yadava, K.C.: *Haryana Ka Itihas,* (in Hindi), Vol. II, 1981.

Zonn, S.V.: *Tropical and Subtropical Soil Science*, Moscow, 1986.

Zubrow, E.B.W. (ed.): *Demographic Anthropology*, New Mexico, 1976.

Zeuner, F.E.: *A History of Domesticated Animals*, London, 1963.

Gazetteers and Reports

Archaeological Survey of India Reports, (Four), 1962-65, Vol. II, 1871.

Census Handbook of Sonipat District, Parts A and B, Chandigarh, 1981.

Census Report of Rohtak District, Chandigarh, 1971.

Census of Sonipat District, Parts XIII A and B, Chandgarh, 1983.

Delhi District Census Hand Book, Part C, Delhi, 1971.

Delhi District Gazetteers, Delhi, 1976.

Ground Water Potential of Jind District of Haryana State, Central Ground Water Board, North-Western Region, Chandigarh, 1981.

Ground Water Potential of Sonipat District, Haryana, Central Ground Water Board, North-Western Region, Chandigarh, 1983.

Haryana District Gazetteers, Rohtak, Simla, 1952.

Haryana District Gazetteers, Rohtak, Chandigarh, 1976.

Haryana District Gazetteers, Karnal, Chandigarh, 1971.

Haryana Government Publication on Art, Chandigarh, 1970.

Hydrological Conditions in The State of Haryana with Possibilities of Ground Water Development, Central Ground Water Board, North-Western Region, Chandigarh, 1983.

Hydrology of Karnal District—Haryana State, 1980-81, Central Water Board, North-Western Region, Chandigarh, 1987.

Punjab District Gazetteers, Delhi, 1910.

Punjab District Gazetteers, Chandigarh, 1961.

Report on the Operations of the Department of Agriuclture, Punjab Official, Annual Publication 1905-47, Issued by the Director of Agriculture.

Technical Report on Archaeological Remains No. 2, Deccan College, Poona, 1961.

Third Regular Settlement Report of Rohtak 1905-10, 1910.

Third Regular Settlement Report of Delhi District 1906-10, 1910.

Theses and Dissertations

Dutt, B.: *Settlements of Painted Grey Ware in Haryana*, Ph.D.,

Thesis (unpublished), Kurukshetra University, Kurukshetra, 1980.

Kumar, Ashok, *History and Archaeology of Kathura Block*, District Rohtak, Haryana, M. Phil Dissertation (unpublished), Maharshi Dayanand University, Rohtak, 1989-90.

Lata, S.: *Archaeology and History of Sampla Block, District Rohtak (Haryana)*, M.Phil., Dissertation (unpublished), Maharshi Dayanand University, Rohtak, 1988.

Singh, J.: *Study of Morphological and Physio-Chemical Characteristics of Soils of Block Ganaur (Sonipat, Haryana) and their Classification*, Ph.D., Thesis (unpublished), Hissar Agriculture University, Hissar, 1971.

Singh, M.: *Archaeological Settlement Pattern of Gurgaon District (Haryana)*, Ph.D., Thesis (unpublished), Kurukshetra University, Kurukshetra, 1990.

Singh, S.: *Archaeology of Sonipat*, M.Phil., Dissertation (unpublished), Kurukshetra University, Kurukshetra, 1984.

Thakran, R.C.: *Sequel to the Indus Civilization in the Greater Indus Valley*, M.Phil., Dissertation (unpublished) Jawaharlal Nehru University, New Delhi, 1981.

Periodicals

American Anthropologist, Chicago

Ancient India, New Delhi.

Ancient Pakistan, Peshawar.

Annual Report of the Archaeological Survey of India, New Delhi.

Anthropology, New Haven.

Antiquity, London.

Archaeological Survey of India Reports, New Delhi.

Bharati, Bulletin of College of Indology, Varanasi.

Bulletin of Bureau of American Ethnology, Perce.

Epigraphia Indica New Delhi.

Geographical Journal, London.

Haryana Studies in History and Culture, Kurukshetra.

Indian Antiquary.

Indian Archaeology—A Review, New Delhi.

Indian Historical Quarterly, Calcutta.

Indian Historical Review, New Delhi.

Journal of American Oriental Society, New Haven.

Journal of Asiatic Society of Bengal, Calcutta.

Journal of Haryana Studies, Kurukshetra.

Journal of Numismatic Society of India, Varanasi.

Kurukshetra University Research Journal, Kurukshetra.

Man and Environment, Poona.

Man in India, Ranchi.

Memoirs of the Archaeological Survey of India, Calcutta.

Man and Environment, Poona

Numismatic Chronicle, 1872.

Pakistan Archaeology–A Review, Peshawar.

The Palaeobotanist, Lucknow.

Proceedings of Indian History Congress, Delhi.

Puratattva, New Delhi.

Studies in History New Series, New Delhi.

Punjab History Conference, Patiala.

Quarterly Review of Historical Studies, Calcutta.

Social Science Probings, New Delhi.

Social Scientist, New Delhi.

Vishveshvranand Indological Journal, Hoshiarpur.

Index